# Communications
# in Computer and Information Science 155

Rory V. O'Connor   Terry Rout
Fergal McCaffery   Alec Dorling (Eds.)

# Software Process Improvement and Capability Determination

11th International Conference, SPICE 2011
Dublin, Ireland, May 30 – June 1, 2011
Proceedings

 Springer

Volume Editors

Rory V. O'Connor
Lero, The Irish Software Engineering Research Centre
School of Computing
Dublin City University
Dublin, Ireland
Email: roconnor@computing.dcu.ie

Terry Rout
Software Quality Institute
Griffith University
Brisbane, QLD, Australia
Email: t.rout@griffith.edu.au

Fergal McCaffery
Regulated Software Research Group
Lero, The Irish Software Engineering Research Centre
Dundalk Institute of Technology
Dundalk, Ireland
Email: fergal.mccaffery@dkit.ie

Alec Dorling
InterSPICE Ltd
Cambridge, UK
E-mail: alec.dorling@interspice.uk.com

ISSN 1865-0929         e-ISSN 1865-0937
ISBN 978-3-642-21232-1         e-ISBN 978-3-642-21233-8
DOI 10.1007/978-3-642-21233-8
Springer Heidelberg Dordrecht London New York

Library of Congress Control Number: Applied for

CR Subject Classification (1998): D.2.9, D.2, K.6, C.2, J.1, K.6.5

© Springer-Verlag Berlin Heidelberg 2011
This work is subject to copyright. All rights are reserved, whether the whole or part of the material is concerned, specifically the rights of translation, reprinting, re-use of illustrations, recitation, broadcasting, reproduction on microfilms or in any other way, and storage in data banks. Duplication of this publication or parts thereof is permitted only under the provisions of the German Copyright Law of September 9, 1965, in its current version, and permission for use must always be obtained from Springer. Violations are liable to prosecution under the German Copyright Law.
The use of general descriptive names, registered names, trademarks, etc. in this publication does not imply, even in the absence of a specific statement, that such names are exempt from the relevant protective laws and regulations and therefore free for general use.

*Typesetting:* Camera-ready by author, data conversion by Scientific Publishing Services, Chennai, India

Printed on acid-free paper

Springer is part of Springer Science+Business Media (www.springer.com)

# Preface

On behalf of the SPICE Organizing Committee we are proud to present the proceedings of the 11th International Conference on Software Process Improvement and Capability dEtermination (SPICE 2011), held in Dublin, Ireland, from May 30 to June 1, 2011.

The SPICE Project was formed in 1993 to support the development of an international standard for software process assessment. The work of the project has eventually led to the finalization of ISO/IEC 15504 – Process Assessment, and its complete publication represented a climax for the work of the project. As part of its charter to provide ongoing publicity and transition support for the emerging standard, the project organized a number of SPICE Workshops and Seminars, with invited speakers drawn from project participants.

These have now evolved to a sustaining set of international conferences with broad participation from academics and industry with a common interest in model-based process improvement. This was the 11th in the series of conferences organized by the SPICE Users Group to increase knowledge and understanding of the International Standard, and of the technique of process assessment.

The conference program featured invited talks, research papers, and industry experience reports on the most relevant topics related to software process assessment and improvement. The technical research papers were selected for presentation following peer review by members of the Program Committee. In addition, a number of tutorials were hosted.

SPICE conferences have a long history of attracting attendees from industry and academia. This confirms that the conference covers topics which are up-to-date, important, and interesting. SPICE 2011 offered a unique forum for industry and academic professionals to discuss their needs and ideas in the area of software process assessment and improvement, and related software quality aspects.

On behalf of the SPICE 2011 conference Organizing Committee, we would like to thank all participants. Firstly all the authors, whose quality work is the essence of the conference, and the members of the Program Committee, who helped us with their expertise and diligence in reviewing all of the submissions. As we all know, producing a conference requires the effort of many individuals. We wish to thank also all the members of our Organizing Committee, whose work and commitment were invaluable.

May 2011

Rory V. O'Connor
Terry Rout
Fergal McCaffery
Alec Dorling

# Organization

## Program Chair

Terry Rout — Griffith University, Australia

## Local Organizing Chair

Rory V. O'Connor — Dublin City University, Ireland and Lero, the Irish Software Engineering Research Centre, Ireland

## Industry Chair

Fergal McCaffery — Dundalk Institute of Technology, Ireland and Lero, the Irish Software Engineering Research Centre, Ireland

## Tutorial Chair

Timo Varkoi — Tampere University of Technology, Finland

## General Chair

Alec Dorling — InterSPICE, UK

## Program Committee

| | |
|---|---|
| Beatrix Barafort | Luxembourg |
| Luigi Buglione | Italy |
| Aileen Cater-Steel | Australia |
| Melanie Cheong | Australia |
| Gerhard Chroust | Austria |
| Francois Coallier | Canada |
| Antonio Coletta | Italy |
| Carol Dekkers | USA |
| Fabrizzio Fabbrini | Italy |
| Mario Fusani | Italy |
| Dennis Goldenson | USA |
| Christiane Gresse von Wangenheim | Brazil |
| Victoria Hailey | Canada |

| | |
|---|---|
| John Horch | USA |
| Linda Ibrahim | USA |
| Ravindra Joshi | India |
| Ho-Won Jung | South Korea |
| Giuseppe Lami | Italy |
| Jean Pierre Legras | France |
| Marion Lepmets | Luxembourg |
| Catriona Mackie | UK |
| Tom McBride | Australia |
| Takeshige Miyoshi | Japan |
| Risto Nevalainen | Finland |
| Hanna Oktaba | Mexico |
| Mark Paulk | USA |
| Alain Renault | Luxembourg |
| Patricia Rodriguez Dapena | Spain |
| Clenio Salviano | Brazil |
| Marty Sanders | Ireland |
| Jean-Martin Simon | France |
| Fritz Stallinger | Austria |
| Robert Treffny | Germany |
| Angela Tuffley | Australia |
| Tanin Uthanayaka | Thailand |
| Han van Loon | Switzerland |
| Timo Varkoi | Finland |

## Local Organizing Committee

| | |
|---|---|
| Paul Brennan | Dundalk Institute of Technology |
| Val Casey | Dundalk Institute of Technology |
| Paul Clarke | Lero Graduate School in Software Engineering |
| Fergal McCaffery | Dundalk Institute of Technology |
| Martin McHugh | Dundalk Institute of Technology |
| Rory V. O'Connor | Dublin City University |
| Fran O'Hara | Inspire Quality Services |
| Sean Rice | BCO Consultants |
| Sivakumar Madhavakuruppu Sudhakar | Dundalk Institute of Technology |
| Murat Yilmaz | Lero Graduate School in Software Engineering |

## Acknowledgments

The local organizers acknowledge the support of the Office of the Vice-President for Research at Dublin City University and Lero, the Irish Software Engineering Research Centre. In addition we acknowledge the support of Science Foundation Ireland in the production of these proceedings.

The conference organizers wish to acknowledge the assistance and support of the SPICE User Group, SPICE 2011 Program Committee and reviewers in contributing to a successful conference.

# Table of Contents

## Process Modelling and Assessment

Using Composition Trees to Model and Compare Software Process ..... 1
  *Lian Wen, David Tuffley, and Terry Rout*

A Modeling View of Process Improvement ........................... 16
  *Clênio F. Salviano*

An Approach to Evaluating Software Process Adaptation ............. 28
  *Paul Clarke and Rory V. O'Connor*

Evaluation of Software Process Assessment Methods – Case Study ..... 42
  *Mohammad Zarour, Alain Abran, and Jean-Marc Desharnais*

## Safety and Security

Functional Safety Extensions to Automotive SPICE According to ISO 26262 ............................................................. 52
  *Per Johannessen, Öjvind Halonen, and Ola Örsmark*

An ISO/IEC 15504 Security Extension ............................... 64
  *Antoni Lluís Mesquida, Antònia Mas, and Esperança Amengual*

## Medi SPICE

Verification & Validation in Medi SPICE ........................... 73
  *M.S. Sivakumar, Valentine Casey, Fergal McCaffery, and Gerry Coleman*

Medical Device Software Development - A Perspective from a Lean Manufacturing Plant ............................................... 84
  *Oisín Cawley, Ita Richardson, and Xiaofeng Wang*

Standalone Software as an Active Medical Device ................... 97
  *Martin McHugh, Fergal McCaffery, and Valentine Casey*

## High Maturity

Assessment of Software Process and Metrics to Support Quantitative Understanding: Experience from an Undefined Task Management Process ........................................................... 108
  *Ayca Tarhan and Onur Demirors*

Methodical Enhancement of Maturity Level: "SPICE" and "SixSigma"
Intertwine.................................................... 121
   *Timo Karasch and Jens Peter Benthaus*

## Implementation and Improvement

Organizational Support for Process Improvement – Results of an
International Survey............................................ 133
   *Marion Lepmets, Eric Ras, and Alain Renault*

Towards a Systemic Maturity Model for Public Software Ecosystems ... 145
   *Angela M. Alves, Marcelo Pessoa, and Clênio F. Salviano*

Linking Software Life Cycle Activities with Product Strategy and
Economics: Extending ISO/IEC 12207 with Product Management Best
Practices...................................................... 157
   *Fritz Stallinger, Robert Neumann, Robert Schossleitner, and
   Rene Zeilinger*

Applying ISO/IEC 12207:2008 with SCRUM and Agile Methods ....... 169
   *Emanuel Irrazabal, Felipe Vásquez, Rafael Díaz, and Javier Garzás*

## Short Papers

Agile and SPICE Capability Levels................................. 181
   *Celestina Bianco*

Application of Lean Principles in Automotive Software Projects........ 186
   *Smitha Bhandary and Shah Quadri*

Experiences Developing TMMi® as a Public Model .................. 190
   *Matthias Rasking*

Experiences from Informal Test Process Assessments in Ireland – Top
10 Findings ................................................... 194
   *Fran O'Hara*

High Levels of Process Capability in CMMI and ISO/IEC 15504 ....... 197
   *Terry Rout*

Process Innovation Reaping Customer Satisfaction .................... 200
   *B. Sridhar and G. Rajesh*

Process Improvement in an R&D&I Center Using *Enterprise SPICE*
and *SPICE* for *Research* Models ..................................... 204
   *Clênio F. Salviano*

Med-Trace ..................................................... 208
   *Fergal McCaffery and Valentine Casey*

Functional Safety – SPICE for Professionals? ...................... 212
   *Bernhard Sechser*

Challenges for Requirements Development: An Industry Perspective .... 217
   *Sandra Kelly, Frank Keenan, and Fergal McCaffery*

Software Engineering Strategies: Aligning Software Process
Improvement with Strategic Goals ................................. 221
   *Reinhold Plösch, Gustav Pomberger, and Fritz Stallinger*

Deploying Lifecycle Profiles for Very Small Entities: An Early Stage
Industry View .................................................... 227
   *Rory V. O'Connor and Claude Y. Laporte*

Past, Present and Future of Process Improvement in Ireland –
An Industry View ................................................. 231
   *Fran O'Hara*

Neural Network Based Effort Prediction Model for Maintenance
Projects ......................................................... 236
   *V. Bharathi and Udaya Shastry*

A Framework of Organizational E-Readiness Impact on E-Procurement
Implementation ................................................... 240
   *Naseebullah, Shuib Bin Basri, P.D.D. Dominic, and*
   *Muhammad Jehangir Khan*

**Author Index** ...................................................... 247

# Using Composition Trees to Model and Compare Software Process

Lian Wen, David Tuffley, and Terry Rout

Software Quality Institute, Griffith University, Brisbane,
Queensland, Australia
{l.wen,d.tuffley,t.rout}@griffith.edu.au

**Abstract.** Software processes described by natural languages are frequently ambiguous and it is usually difficult to compare the similarity and difference between one process defined in one standard and its counterpart defined in another standard. This paper proposes Composition Tree (CT) as a graphic language to model software process based on its purpose and expected outcomes. CT is a formal graphic notation originally designed for modeling component based software system. This paper demonstrates that CT can be a powerful notation to give a clear and unambiguous description of a software process as well. This paper also investigates an algorithm which can compare two CT-modeled processes and provide an intuitive view called a Comparison Composition Tree (CCT) to highlight the differences and similarities between the two processes.

**Keywords:** Software Process, Behavior Engineering, Composition Tree, Process Reference Model.

## 1 Introduction

Process models are abstract representations of a process architecture, design or definition [6]. They are abstractions, not direct representations of reality. The language that people use when developing these abstractions, these process models, is prone to ambiguity due to the fallible way in which people use language. George Box famously observed that all models are wrong but some are useful [2]. Models are simplifications of reality and in the process of simplifying, essential information might be left out resulting in ambiguity. Even with ambiguity, models can be useful, but this ambiguity points to a need for a way to reduce or eliminate it. Such a way might be found in a formal method such as Behavior Engineering.

In seeking a solution to the problem of ambiguity in process modeling, one sees a similar problem with the requirement specifications for software systems. Ambiguous language, incomplete descriptions, repetition and redundancies in the way specifications are expressed inevitably leads to sub-optimal project outcomes (systems that do not meet the user's needs). Behavior Engineering [3] successfully addresses the problems faced by software developers seeking to translate a set of user requirements into a complete and consistent requirements specification.

Behavior Engineering uses a formally-grounded graphical notation with the capability to represent a wide range of system behaviors in unambiguous terms. Its strength is its ability to accommodate complexity and detail, ease of use, and in particular for this project its ability to expose defects.

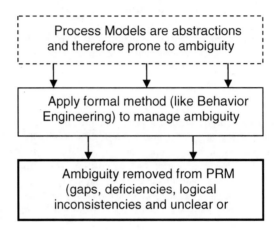

**Fig. 1.** Use formal method to remove ambiguity from abstract model

Previous research indicates that BE notations can be useful verification tools for process modeling [13]. This paper refines this concept by proposing a detailed scheme to model a software process based on its purpose and process outcomes in a Composition Tree (CT) [5], which is one of the key parts of the BE notations. The graphic version of a process model is more intuitive, less ambiguous and easier to verify than the original natural language described process.

With the quick development and diversity of software standards for different domains [12], systematic methods of comparison of software processes are crucial for process analysis, understanding and evolution [10]. However, it is difficult to consistently, systemically and automatically compare two processes if they are described in natural languages.

This paper proposes a formal method which can compare two processes when they are modeled in CTs. This method is based on a precisely defined tree merging algorithm [14]; therefore it can be automated.

The proposed comparison method generates a Comparison Composition Tree (CCT) that explicitly shows the difference and similarity of the two compared processes. In particular, the CCT highlights the difference in a way that is easy to read and understand, so it can be very useful for people to study the evolution of processes.

The paper is organized in the following way: Section 2 and Section 3 provide necessary background information about Process Models and Composition Trees respectively; Section 4 introduces the method to use a CT to model a process; Section 5 describes the algorithm to compare two CTs; Section 6 demonstrates this comparison method through a case study that compares processes for configuration management in two different standards; finally, a brief conclusion is given in Section 7.

## 2 Software Process Models

Feiler and Humphrey [6] define a process model as an *abstract representation of a process architecture, design or definition*. Process models in this broad sense can be seen as process elements at an architectural, design and definitions level. The abstraction inherent in process models serves to capture and represent the essential nature of processes. Any representation of the process can be said to be a process model. Process models can be analyzed, validated, and if enactable can simulate the modeled process [6].

Scacchi [11] distinguishes software process models from software lifecycle models. The former are descriptive or prescriptive characterizations of how software is developed, whereas the latter *represent a networked sequence of activities, objects, transformations, and events that embody strategies for accomplishing software evolution* [11]. This definition is not inconsistent with that of Feiler and Humphrey [6] discussed above. Process models are useful for developing more precise and formalized descriptions of software life cycle activities, using a rich notation, syntax, or semantics, often suitable for computational processing [11]. This idea lends support for the use of Behavior Engineering [4] and its notation that might be accurately described as *rich notation, syntax, or semantics* to develop Process Reference Models.

ISO/IEC 24774:2007 - *Software and systems engineering -- Life cycle management -- Guidelines for process description* [9] outlines a standard format for any process reference model, including those intended for process implementation and process assessment. This general purpose standard outlines the elements used to describe a process; title, purpose statement, outcomes, activities and tasks.

- The **title** conveys the scope of the process as a whole, expressed as a short noun phrase that summarize the scope of the process, identify the principal concern of the process, and distinguishes it from other processes within the scope of a process model.
- The **purpose** describes the goal of performing the process. It is expressed as a high level goal for performing the process, preferably stated in a single sentence. The implementation of the process should provide measurable, tangible benefits to the stakeholders through the expected outcomes.
- The **outcomes** express the observable results expected from the successful performance of the process. Outcomes are expressed in terms of a positive, observable objective or benefit. The list of outcomes associated with a process shall be prefaced by the text, 'As a result of successful implementation of this process:' The outcomes should be no longer than two lines of text, about twenty words. The number of outcomes for a process should fall within the range 3 to 7. Outcomes should express a single result. The use of the word 'and' or 'and/or' to conjoin clauses should be avoided. Outcomes should be written so that it should not require the implementation of a process at any capability level higher than 1 to achieve all of the outcomes, considered as a group.
- The **activities** are a list of actions that may be used to achieve the outcomes. Each activity may be further elaborated as a grouping of related lower level actions;
- The **tasks** are specific actions that may be performed to achieve an activity. Multiple related tasks are often grouped within an activity.

ISO/IEC 24774:2007 [9] makes it clear that the outcomes should not go beyond what is stated in the purpose. There should be no capability level issues expressed in the outcomes.

Secondly the outcomes must address all of the issues that are apparent in the purpose statement. Nothing should be missed. The outcomes must therefore be necessary and sufficient to satisfy the purpose.

## 3 Composition Trees

A Composition Tree is originally used to describe the composition of a component based software intensive system [5]. It provides useful summary information including states, attributes and relationships about the system and other entities under the system.

Similar to the way of constructing a Behaviour Tree from the functional requirements [4], A Composition Tree can also be constructed through translating the individual functional requirements one by one. In this section, a small sized case study is used to explain both the notations of Composition Trees and also the procedure to build a Composition Tree from the functional requirements. In order to help readers to capture the concepts quickly, the introduction is more intuitive but less formal. A formal and more complete description of Composition Trees can be found at the Behaviour Engineering website [1].

**Case study:** a CAR system:

- R1: The car can only be started if it is in the park state when the driver inserts the key in the ignition and turns it on.
- R2: A dashboard light remains on if the driver's seatbelt is not fastened when the driver is seated and the ignition is on.
- R3: If the handbrake is on when the ignition is on, the brake-light turns on.
- R4: The security alarm is on when the car is locked, and if anyone tries to break in by breaking a window or forcing a door the alarm will sound.
- R5: When the driver, on approaching the car, presses the key-button it unlocks the door and turns the security alarm off.
- R6: When the car is unlocked the driver may get in and put the car into the park state.

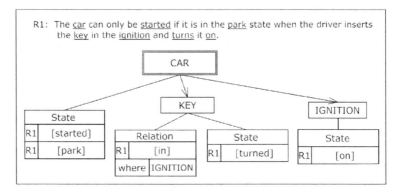

**Fig. 2.** The Composition Tree (CT) generated from translating R1

Fig. 2 is the Composition Tree (CT) created from Requirement 1 (R1). This diagram shows the following information:

- There are two components KEY and IGNITION[1] under the system component CAR, which is drawn in doubled line.
- The CAR system has two different states "started" or "park".
- The IGNITION has one state "on" and the KEY has one state "turned".
- The KEY has one relation which is in the IGNITION.
- The requirement tag R1 helps people to trace the information in the diagram back to the original requirement.

Please note that all the compositional information shown in Fig. 2 is faithfully translated from the original requirement including most of the terminology. However the information presented in the graph is more precise and less ambiguous.

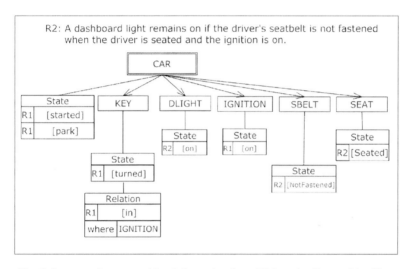

**Fig. 3.** Integrate the composition information from R2 into the Composition Tree

Fig. 3 integrates the compositional information in R2 into the CT. The new tree may also be called an Integrated Composition Tree (ICT). However, in order to avoid unnecessary confusion, we will only use Composition Tree (CT) in this paper. The new CT shows the following addition information:

- There are three more components DLIGHT (dashboard light), SBELT (seatbelt) and SEAT[2].
- The DLIGHT has state "on", the SBELT has state "notfastened" and the SEAT has state "seated".

---

[1] Some researchers may also model DRIVER as one component in the CAR system. However, this paper considers that the driver is something external to the system.
[2] The SEAT component is not explicitly mentioned in R2, but based on the common understanding of a car, it is a reasonable interpretation of R2.

A composition tree only captures the static compositional information of individual component. The dynamic, logical and cause and effect relationship between components is captured in Behavior Trees [4].

Based on the same process, an entire CT can be generated through integrating all the requirements as in Fig. 4.

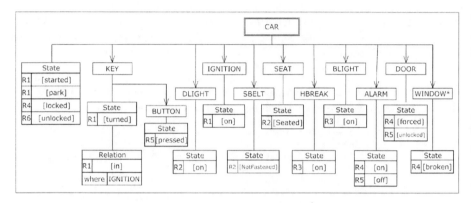

**Fig. 4.** The complete Integrated Composition Tree of the CAR system

Fig. 4 shows the complete list of the components in the CAR system, and the expected states and relations for each component. All the information is directly translated from the user requirements. The advantages of the CT over the natural language described functional requirement are:

- All information is integrated together so it is easy to indentify the requirement defects. For example, the CT shows that many components have only one state. It's obvious that the original requirements are not completed because each component should have at least two different states.
- A CT arranges the information about one component in one place. It will be easier for people to design and implement the components than the original requirements with the information of one component may be scattered all around the requirements.
- The more specific graphic notation is less ambiguous than the more flexible natural language.
- A CT removes the entire alias so it will use a consistent vocabulary for the system.

There are more discussions about the advantage of using CT in Software and System Engineering in elsewhere [5], which will not be repeated here. The focus of this paper is to investigate using CTs for Process Modeling.

## 4 Using the Composition Tree Approach to Model Process

According to ISO/IEC 24774:2007 [9], the standard elements to describe a process include the title, the purpose, outcomes, activities and tasks. Apart from the title, which is only the name of a process, purpose and outcomes are more static elements, so they may be more suitable to be modeled by composition trees.

This paper proposes a way to use a Composition Tree (CT) to model a software process based on its purpose and outcomes, which are usually documented in natural languages.

The way to construct a CT from the process purpose and outcomes includes the following steps:

1. Read through the purpose and outcomes, and make a complete and consistent list of nouns and acronyms, which are usually components or attributes of components.
2. Starting from the process purpose state, identify the components and their state and draw the initial CT.
3. Read each outcome one by one, to identify the components, states, relationship and attributes and then integrate the information in the CT.

This paper uses the Configuration Management process defined in ISO/IEC 12207:2000 [8] to explain the process.

**Process Name:** Software configuration management.

**Process Purpose:** *The purpose of the Configuration management process is to establish and maintain the integrity of the work products/items of a process or project and make them available to concerned parties.*

**Process Outcomes:**
1. *a configuration management strategy is developed;*
2. *items generated by the process or project are identified, defined and baselined;*
3. *modifications and releases of the items are controlled;*
4. *modifications and releases are made available to affected parties;*
5. *the status of the items and modifications are recorded and reported;*
6. *the completeness and consistency of the items is ensured; and*
7. *storage, handling and delivery of the items are controlled.*

In order to model this process, the first step is to identify and list all the components:

**CMP:** Software Configuration Management Process
**WPI:** Work product or item
**CPT:** Concerned Party
**CMS:** Configuration Management Strategy

The second step is to translate the process purpose into a composition tree as below: Fig. 5 shows that there are two different types of components under CMP, WPI and CPT; the "*" sign indicates that the component may have more than one instance. The WPI has an attribute called integrity and the integrity needs to be established and maintained. There is also a relationship for WPI; the relationship is that WPIs should be available to CPTs. The tag "P" in each box means this piece of compositional information is translated from the purpose of the process.

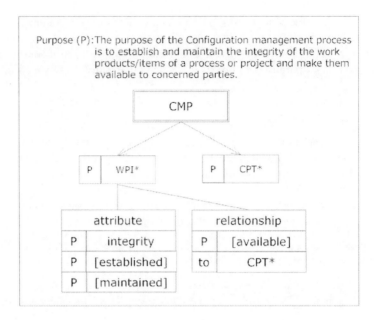

**Fig. 5.** The CT constructed from the purpose of the Configuration Management process

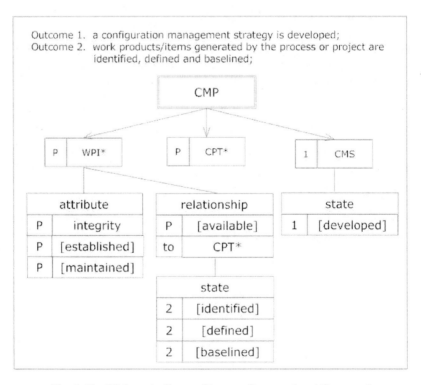

**Fig. 6.** The CT from the Process Purpose, Outcome 1 and Outcome 2

The next step is to translate the outcomes one by one and integrate the compositional information into the CT. Fig. 6 shows the CT after outcome 1 and outcome 2 have been integrated. The compositions of a component include attributes, relationship and states; the three different compositions can be drawn directly under the component or one under another as in Fig.6 (the states of WPI are drawn under the relationship of WPI). There is no semantic difference, the variation is only for an easy to read layout.

Eventually, when all the outcomes are translated and integrated into the CT, it will be a complete CT showing the purpose and the expected outcomes as in Fig. 7.

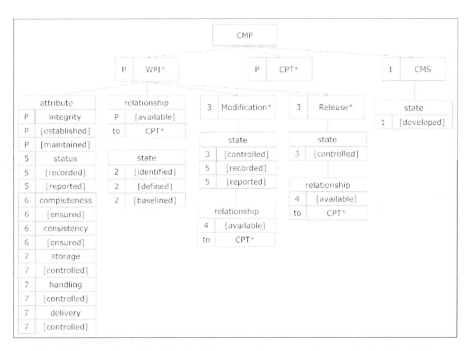

**Fig. 7.** The complete CT of the Configuration Management Process defined in 12207

Fig. 7 shows the Configuration Management Process in a CT. It has a few advantages over the initial natural language description:

1. All information is integrated in one graph, so the relationships between different parts become visible.
2. The information of each component is arranged in one place so it will be easier to retrieve. For example, from Fig. 7, it is easy to find that WPI has the following attributes: integrity, status, completeness, consistency, storage, handling and delivery.

Generally, the graphical version of the process may have less ambiguity, may be easier to understand and easier for people to identify process defects.

## 5 The Algorithm to Compare Two CTs

The previous section has explained how to use a CT to model a software process.

This section will introduce an algorithm which is to compare two CTs of two similar processes and identify the differences.

The comparison implements a label matching tree merging algorithm which has been used for comparing different versions of Behavior Trees [14]. However, as the notation of a composition tree is different from a behavior tree, there are some differences in the merging trees.

A critical task in tree merging algorithm is to identify the matching nodes. For CTs, the way to identify the same nodes is based on the name of component, state, etc. Therefore, before applying the merging algorithm, the first step is to identify the same component and/or same state which may be called by different names in the two compared trees and to establish a mapping between them. For example in Fig. 7, the component of work product and work item is called WPI but the same component may be called WP in another CT. In this situation, a mapping table is required.

The second step is to compare and merge the two trees. To simplify the discussion, we may call the first tree as the old tree and then the second tree as the new tree. In this way, the comparison procedure will create a merged tree that is called a Comparison Composition Tree (CCT). A CCT shows all the information of both trees and also highlights the difference in an easy to read way. To achieve this purpose, a display style convention is used in this paper as in Fig. 8.

Under this display style convention, in a CCT, a piece of information which exists in both the old tree and the new tree is called unchanged and will be drawn in normal style; a piece of information if only exists in the old tree will be called old and will be drawn in dotted lines; a piece of information if only exists in the new tree will be called new and will be drawn in bolded lines.

**Fig. 8.** The display style convention for a CCT

Now we will use a simple abstract example to explain the tree merging algorithm. Suppose that $T_1$ and $T_2$, shown in Fig. 9, are the old CT and the new CT respectively.

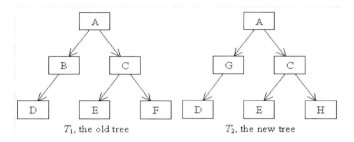

**Fig. 9.** The old CT $T_1$ and the new CT $T_2$

To compare $T_1$ and $T_2$ and generate the CCT, we use the following algorithm:

1. Start the comparison from the root nodes (in this example, node A). Because the root node exists in both trees, it is created in the CCT as an unchanged node.
2. Find the compared node's child-node set in both trees. (In this example, the child-node set in the old tree is {B, C} and the child-node set in the new tree is {G, C}.
3. If a node exists in the old tree's child-node set but not in the new tree's child-node set, this node will be marked as an old node in the CCT. (In this example, B is such a node)
4. In the old tree, the sub trees under the old node will be generated in the CCT as old. (In this example, the node D under node B in $T_1$ is such a case)
5. If a node exists in the new tree's child-node set but not in the old tree's child node set, this node will be created in the CCT as a new node. (In the example, G is such a node)
6. In the new tree, the sub trees under the new node will be generated in the CCT as new. (In this example, the node D under node G in $T_2$ is such a case)
7. If a node exists in the child-node sets of both trees, it will be generated in the CCT as an unchanged node. (In the example, the node C is such a case)
8. An unchanged node will be a new comparison node and the algorithm will go back recursively to step 2.

The CCT $T_c$ produced from $T_1$ and $T_2$ is shown in Fig. 10, following the style convention used in Fig. 8.

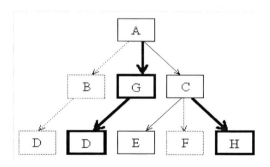

**Fig. 10.** The Comparison Composition Tree $T_c$

## 6 Case Study

The previous section uses two abstract trees to explain the tree merging algorithm. This section will apply this algorithm on a case study based on real processes.

Section 4 has used a CT to model the Configuration Management process defined in ISO/IEC 12207 [8] (see Fig. 7). A similar Configuration Management process has also been defined in ISO/IEC 15288: 2002 [7] for system engineering. The next step is to model the Configuration Management process from 15288 and compare it with that from 12207.

The purpose and outcomes of the Configuration Management process in 15288 are quoted below:

**Purpose (P):** *The purpose of the Configuration Management Process is to establish and maintain the integrity of all identified outputs of a project or process and make them available to concerned parties.*

**Outcomes:**
a) *A configuration management strategy is defined.*
b) *Items requiring configuration management are defined.*
c) *Configuration baselines are established.*
d) *Changes to items under configuration management are controlled.*
e) *The configuration of released items is controlled.*
f) *The status of items under configuration management is made available throughout the life cycle.*

Based on the purpose and outcome, following components are identified:
**CMP:** Configuration Management Process
**ITM:** identified output / item
**CPT:** Concerned Party
**CMS:** Configuration Management Strategy
**CFB:** Configuration Baseline

The CT of the Configuration Management process defined in 15228 is drawn in Fig. 11.

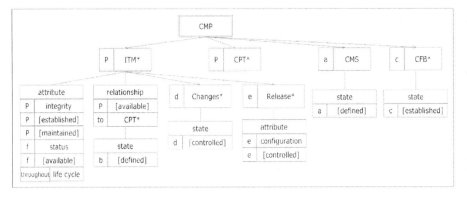

**Fig. 11.** The CT of the Configuration Management process defined in 15228

Considering that the CT in Fig. 11 is the old tree and the CT in Fig. 7 is the new tree, the comparing process will merge them together to generate a CCT. The first step is to create a list of mapping terms used in the two CTs. The list is shown in Table 1. After the mapping table has been established, the two CTs can be compared and merged into one CCT as in Fig. 12.

**Table 1.** The mapping terms in the two processes

| # | 15228 | 12207 | Comments |
|---|-------|-------|----------|
| 1 | ITM | WPI | Work product is a common term in software process, while item is a more general name. |
| 2 | Change | Modification | They are sub component order ITM/WPI |

Fig. 12 shows the similarities and differences of the Configuration Management Process defined in both 12207 and 15228. There are a number of interesting things can be found in this CCT:

- There is component called Configuration Baseline (CFB) in 15228 but no such component mentioned in 12207. However, 12207 has mentioned that the WPI should be baselined.
- 12207 requires a configuration management strategy (CMS) to be developed, but 15228 asks a CMS to be defined. Can we assume that "developed" and "defined" mean the same thing regarding to a CMS?
- Both 12207 and 15228 have some requirements on the release of work product or items. However, 12207 asks the releases to be controlled and to be available to the concerned parties, while 15228 asks only the configuration of releases to be controlled.
- For the modifications (or changes) of a WPI/ITM, both standards ask them to be controlled, but 12207 also ask them to be recorded, reported and available to the concerned parties.
- Both standards ask WPI/ITMs to be available to the CPTs and defined, but 12207 has mentioned that the WPI/ITMs should also be identified and baselined.
- Both standards ask the integrity of WPI/ITM* to be established and maintained.
- 12207 asks the status of WPI/ITMs should be recorded and reported, but 15228 only ask it to be available (throughout life cycle).
- 12207 has some requirements regarding to the completeness, consistency, storage, handling and delivery of WPIs but there is no corresponding requirements explicitly mentioned in 15228.

Fig. 12 demonstrates how compositions can be used to highlight the similarity and difference between two processes. The advantage of a CCT is that it shows the similarity and difference between in a clear, complete and unambiguous way. This information can be helpful for people who are developing new process models. It also offers a tool of considerable strength to support the harmonization of process models drawn from different domains, but where the essential purpose of the process is the same.

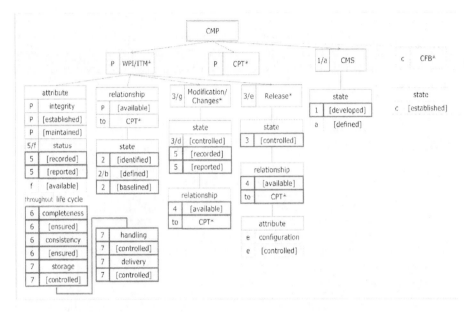

**Fig. 12.** The CCT between the Configuration Management Process of 12207 and 15228

## 7 Conclusions

The paper has proposed the formal graphic notation known as a Composition Tree (CT) as a viable way to verify processes and process model[5]. The graphic notation is intuitive and unambiguous and makes it easier to define semantics.

The paper extends the use of CTs by applying a tree merging algorithm [14] that compares the CTs of two similar processes to generate a Comparison Composition Tree (CCT). This displays the difference and similarity of two compared processes in a clear and user friendly way.

The results to date are promising. Even still at a preliminary stage, further researches including formal semantics of the process CT, automation tools and large case studies are on the way, the proposed method can be useful for people to study software processes as well as to design new processes. At this stage, application of the approach has been limited to the comparison of processes described using the same basic modeling approach (Process Reference Models). It has previously been demonstrated [13] that the approach can be employed to analyse other types of process model; it should therefore, be possible to generate CCT representations to compare models of the same process generated using different modeling approaches – for example, to compare a Process Reference Model with an Implementation Model and a Process Assessment Model addressing the same process, thus enabling validation across multiple models. This capability offers considerable benefits for developers of process models, whether such models are for intended for standardization, instruction or improvement purposes. Future research seems definitely warranted, and should focus on exploring the full capabilities of CT in relation to these issues.

# References

1. Behavior Engineering Web Site, http://www.behaviorengineering.org/
2. Box, G.E.P.: Robustness in the strategy of scientific model building. In: Launer, R.L., Wilkinson, G.N. (eds.) Robustness in Statistics. Academic Press, New York (1979)
3. Dromey, R.G.: Climbing Over the 'No Silver Bullet' Brick Wall. IEEE Software 23(2), 118–120 (2006)
4. Dromey, R.G.: Formalizing the Transition from Requirements to Design. In: Liu, Z., He, J. (eds.) Mathematical Frameworks for Component Software, Models for Analysis and Synthesis, ch. 6, pp. 173–206. World Scientific, Singapore (2006), ISBN 981-270-017-X
5. Dromey, R.G.: System Composition: Constructive Support for the Analysis and Design of Large Systems. In: SETE-2005, Systems Engineering/Test and Evaluation Conference, Brisbane, Australia (2005)
6. Feiler, P.H., Humphrey, W.S.: Software Process Development and Enactment. Software Engineering Institute, Pittsburgh, CMU/SEY-92-TR-04, p. 11 (1992)
7. ISO/IEC 15288:2002. Information technology - System engineering – System life cycle process (2002)
8. ISO/IEC 12207:2008 – Information technology – Software engineering – Software life cycle processes (2008)
9. ISO/IEC TR 24774. Software and systems engineering – Life cycle management – Guidelines for process description (2007)
10. Podorozhny, R.M., Perry, D.E., Osterweil, L.J.: Artifact-based functional comparison of software processes. In: 4th International Workshop on Software Process Simulation and Modeling, May 2003, pp. V.29.1–V.29.10 (2003)
11. Scacchi, W.: Process Models in Software Engineering. Encyclopedia of Software Engineering. In: Marciniak, J.J. (ed.) Encyclopedia of Software Engineering, 2nd edn. John Wiley and Sons, Inc., New York (2001)
12. Sheard, S.A.: The frameworks quagmire, a brief look. In: Proceedings of the 7th Annual International INCOSE, Symposium (INCOSE 1997) (1997)
13. Tuffley, D., Rout, T.: Behavior Engineering as Process Model Verification Tool. In: The proceedings of the 10th International SPICE conference (2010)
14. Wen, L., Dromey, R.G.: From Requirements Change to Design Change: A Formal Path. In: Proceedings of the 2nd IEEE International Conference on Software Engineering and Formal Methods, pp. 104–113 (2004)

# A Modeling View of Process Improvement

Clênio F. Salviano

CTI: Centro de Tecnologia da Informação Renato Archer
Rodovia D. Pedro I, km 143.6, CEP 13069-90, Campinas, SP, Brazil
Clenio.Salviano@{cti.gov.br,gmail.com}

**Abstract.** As a consequence of Software Process Improvement success there are forces that urge for further evolution. One force is the need for eliciting and refining underlying SPI principles. This article introduces a modeling view of process and process improvement with three types of process models (Process Capability Profile, Process Enactment Description and Process Performance Indicator) and an example on a process improvement cycle. This modeling view improves the integrated understanding of what we want, what is the current status, what we can do and what we are doing for improvement during a cycle. This modeling view is then used as a basis for introducing Modeling driven (Knowledge Working) Process Improvement as an evolution of current Model-based (Software, Systems and Services) Process Improvement.

**Keywords:** SPI, Process Modeling, PRO2PI Methodology, SPICE, CMMI.

## 1 Introduction

Software Process Improvement (SPI) has been a successful methodology for the necessary improvement of software development. SPI started about twenty five years ago with the development and usage of SW-CMM (Capability Maturity Model for Software) [1] and SPICE (Software Process Assessment and Capability dEtermination) [2] models. There have been evolutions on SPI, including its generalization from software to software, system and services, and the movement from models to framework of models. Now a days, CMMI (Capability Maturity Model Integration) [3] and ISO/IEC 15504 (SPICE) [4] (and its ongoing revision towards ISO/IEC 33000 Series [5]) are the most dominants frameworks for models and SPI. SPI methodology has been based in pre-defined models of best practices. Hence, it can be identified as "Model-based process improvement".

As a consequence of SPI success, there are forces that urge for further SPI evolution. A previous article identified seven groups of these forces [6]. One of these forces is the need for eliciting and refining underlying principles of SPI. Card states that "different approaches [for current SPI] are considered competitors, even though they are all based on very similar concepts and techniques. The current packaging obscures the underlying principles" [16]. ISO/IEC 15504 (SPICE) and the recent SPI Manifesto [7] advanced these underlying principles. SPICE provides requirements for process assessments models and documented process assessments processes. The SPI Manifest elicits the true values and principles of SPI, showing the need to emphasize improvements over conformance to pre-defined models.

A challenge is how to evolve SPI in order to balance these forces. This article proposes modeling as the main reference to conduct this evolution. It proposes "Modeling driven Process Improvement" as an evolution of "Model-based Process Improvement". This article is organized as follows. This first section provides an introduction to the article. The second section provides an overview of model, metamodel, modeling and chain of models concepts. The third and fourth sections introduce a modeling view of process and process improvement with three types of process models and an example on theirs usage in a process improvement cycle. The fifth section analyses how two well established SPI frameworks (CMMI and ISO/IEC 15504) cover these three types of process models. The sixth section uses this modeling view as a basis for introducing Modeling driven (Knowledge Working) Process Improvement as an evolution of current Model-based (Software, Systems and Services) Process Improvement. Finally, the seventh section concludes the article.

The content of this article is part of an ongoing Research, Development and Innovation (R&D&I) effort on process improvement by *CTI Renato Archer* and its partners since 1999. CTI is a Brazilian Information Technology R&D&I Center (www.cti.gov.br). This R&D&I effort has been conducted with many cycles of industry demand, exploration, application and consolidation following the industry-as-laboratory research approach proposed by Potts [8] as the R&D&I methodology.

## 2 Model, Metamodel, Modeling and Chain of Models

This section presents basic concepts related with model, modeling and chain of models as references for the next sections. Bézivin, Favre and other authors [9] [10] [11] define model as "a simplification of a system built with an intended goal in mind" and complete this definition with "a model represents certain specific aspects of a system and only these aspects". Therefore there are three elements in a model: the system, the intended goal and the aspects. An intended goal of a model is to be able to answer questions in place of the actual system [9]. In a more precise statement a model follows the Limited Substitutability Principle: "The purpose of a model is always to be able to answer some specific sets of questions in place of the system, exactly in the same way the system itself would have answered similar questions" [9].

The correspondence between a system and a model is precisely defined by a metamodel. Each metamodel is used to specify which particular "aspect" of a system should be considered to constitute the model. A metamodel defines a consensual agreement on how elements of a system should be selected to produce a given model [9]. A metamodel is not a model of a model. Rather, a metamodel is a model of a language of models [10].

A model can be used as a specification model, that represents a system to be built, or as a descriptive model that describes an existing system. New systems are produced from specification models. Descriptive models are produced from existing systems [10]. There is also the notion of co-evolution of model and system [10], where both model and system are in constant evolution and each version of the model is either a specification or a descriptive model.

Rothenberg indicates the meaning of modeling in the broadest sense as "the cost-effective [development and] use of something [, a model,] in place of something else for some cognitive purpose" [11]. Bézivin complements this definition stating that "modeling is essential to human activity because every action is preceded by the construct (implicit or explicit) of a model [or a set of models]" [9].

Be a model is not an intrinsic characteristic of an artifact. Rather it is a relationship between two artifacts. The same artifact can be a model of a system in one relationship and a system being modeled in another. Actually there is a chain of models, where a model in one relationship became a system in another, and so on. Peirce explored this concept of chain of models in his semiotics. Peirce´s semiotics provides a scientific basis for modeling [12]. So, Bézivin´s statement can be rephrased as "modeling is essential to knowledge working process improvement, because every human action is preceded by the construct (implicit or explicit) of chains of specification and descriptive models".

## 3 Process as a Model and Types of Process Models

In SPI there are two popular definitions of (software) process. One is "process is what people do" [1] and the other is "a set of interrelated (or interacting) activities, which transform inputs into outputs, to achieve a given purpose" [3] [4]. From a modeling perspective, they complement each other. The first one defines process by the modeled system and the second one defines process by a model of that system. A proposed definition is process is a model of what people as a set of interrelated or interacting activities which transform inputs into outputs, to achieve a given purpose.

There is a trend to generalize SPI from software process to software, systems, services and other domains processes. A proposal is to use knowledge worker for this generalization. Knowledge worker is used in the sense defined first by Drucker [13] as "anyone who works for a living at the tasks of developing or using knowledge. Knowledge working is the activity of the knowledge worker. Knowledge workers have high degrees of expertise, education, or experience, and the primary purpose of their jobs involves the creation, distribution, or application of knowledge" [13].

Process however stills not enough. So, process became a system and modeling is used again to produce process models. From our experience, three types of process models are more important for SPI. Each one represents a process under a different perspective. Each one, based on that perspective, uses a set of elements and allows answers for some specific set of questions in a close and useful enough "way the system [in this case, the process] itself would have answered similar questions". Figure 1 illustrates process as a model and types of process models.

The twelve people icons in the bottom part of Figure 1 illustrate knowledge workers working. (Knowledge working) Process is a model of "what knowledge workers do and think" "for an objective, transforming inputs into outputs" built with an intention to "improve" the work. The work is the system in the model-system relationship. The cloud with selected and organized view of people icons, including input and outputs, illustrate process as a model.

The three pairs of graphic icons in the upper part of Figure 1 illustrate three types of process models that are more important during a process improvement cycle. Each one represents a process under a different perspective (or dimension). Each one, based on that perspective, uses a set of elements. Process is a model in the model/system relationship in the bottom part of Figure 1 and the same process is a system in each one of the three model-system relationship in the upper part.

The three icons in the very upper part of Figure 1 illustrate three more models in these chains of model-system relationship. Each icon is a model of a type of process model. Therefore there are seven models in Figure 1. All of them are models of what knowledge workers do and think for a living.

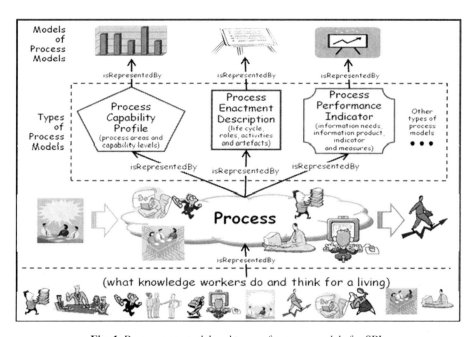

**Fig. 1.** Process as a model and types of process models for SPI

The three icons in the very upper part of Figure 1 illustrate three more models in these chains of model-system relationship. Each icon is a model of a type of process model. Therefore there are seven models in Figure 1. All of them are models of what knowledge workers do and think for a living.

A Process Capability Profile model is structured with process (or process areas) and capability levels, as defined, for example, by ISO/IEC 15504-5 Process Assessment Model [4]. A Process Enactment Description model is structured with life cycle, roles, activities and artifacts, as defined, for example, by Software Process Engineering Metamodel (SPEM). A Process Performance Indicator model is structured with information needs, information product, indicator and measures, as defined, for example, by Practical Software and System Measurement (PSM) and ISO/IEC 15939 Software Measurement Process.

Suppose we don´t know the process of a given organization. We know, however, that Maturity Level 3 of the Exemplar Organizational Maturity Model defined in ISO/IEC 15504-7 (15504-ML3) (or Maturity Level 3 of CMMI-DEV model – CMMI-ML3) is a

Process Capability Profile model of this process. What questions can we answer for this process? For example, the question "Can we have a good confidence that this organization will deliver functional software, on time?" can be answered with "Yes" because 15504-ML3 (or CMMI-ML3) allows this answer. What questions we cannot answer for this process? For example, the question "Will this organization deliver incremental versions during the development or everything at the end?" cannot be answered because 15504-ML3 does not allow answer for this question. The answer depends on the life cycle model.

Suppose we also know, however, a Process Enactment Description model of this process and it says that the process uses an incremental life cycle. We then can answer "Incremental versions" for this question. What questions we still cannot answer for this process? For example, a question about which level of quality (in terms of percentage of serious faults) should we expect for each delivery, cannot be answered because neither one of the two previous models allow an answer for this question. Suppose we also know, however, a Process Performance Indicator model of this process and it says that "98% of all delivery software systems have less than 2 shipped defects per thousand of source lines of code". Then we can answer this question because this model allows this answer.

## 4   A Process Modeling View of Process Improvement

This section presents examples of a process modeling view of a process improvement cycle. Figure 2 provides a simplified and high density illustration with process modeling view´s snapshots of a SPI cycle with the three types of process models.

In Figure 2 the bigger gray arrow shows the flow of a SPI cycle. The smaller gray arrows show sub cycles of the implementation phase of the SPI cycle. The three small icons and the four clouds in the middle represent versions of the current or future process. Each question and its correspondent geometric form represent either a descriptive model (*D Model*) of the current process or a specification model (*S Model*) of a future process. The position of each model and the index of the modeled process indicate roughly where the correspondent modeling is performed in the SPI cycle.

In the beginning of what is going to be a SPI cycle, a manager wants to know what is going on in a software development department. As answer, she got the impression that "most projects seem to be very late" and concludes that this is not good for business. From process modeling point of view, she got "most projects seem to be very late" as a vague Process Performance Indicator descriptive model of the current process. As a second step, she wants to know what performance indication could be feasible and good for them. After some inquiries, she concludes that "90% of projects, +/- 10% on time" (at least 90% of all relevant software development projects on time with an accepted interval up to 10% delay or anticipating) is a satisfying and feasible indicator. If, in the future, an improved process results in that indicator, the business will be better. From process modeling point of view, this indicator is a (more precise) Process Performance Indicator specification model for a future process.

A Modeling View of Process Improvement    21

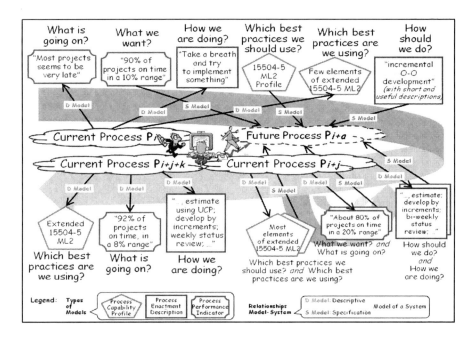

**Fig. 2.** Snapshots of process modeling view of a SPI cycle

As a third step, she wants to know how they are developing software projects. She got the impression that there is no standard process to develop software projects. The process is improvised for each project and there is no planning for it. She concludes that the current process is something as "take a breath and try to implement something". From process modeling point of view, she got a Process Enacting Description descriptive model of the current process.

As a fourth step, she wants to know which best practices can help. After some inquiries, she concludes that ISO/IEC 15504-5 or CMMI-DEV model of best practices could help. More precisely, the Maturity Level 2 of the Exemplar Organizational Maturity Model in ISO/IEC 15504-7 (15504 ML2) (or Maturity Level 2 of CMMI-DEV model) could be used as reference for feasible and useful best practices. If, in the future, an improved process implements the set of best practices of 15504 ML2, the business will be better. From process modeling point of view, the 15504 ML2 is a Process Capability Profile specification model for a future process.

As a fifth step, she wants to know how they are using the best practices from 15504 ML2. To answer that question, she contracts a process assessment. This process assessment results that they implement just few of these 15504 ML2 best practices. The process assessment also identifies some other few best practices that have been performed without be part of the 15504 ML2. The profile composed of few elements of 15504 ML2 and the other few best practices is, from process modeling point of view, a Process Capability Profile descriptive model of the current process.

As a sixth step, she wants to know requirements for process enactment descriptions. She concludes that these descriptions should be "short and useful" and its first version is "incremental O-O development". From process modeling point of view, the "short and useful" are requirements for and "incremental O-O development" is a Process Enactment Description specification model for a future process.

At that point, from process modeling point of view, the organization has three descriptive models of its current process and three specification models for a future process. These models are established to provide a basis for improving the process.

Then, sub-cycles of process analysis and process change proposal, analyses, implementation and revision are performed. At same point a Process Engineering Group analyzes the current process and together with a Process Steering group decide to improve the current process description to incorporate an orientation to perform bi-weekly status review of the software development project. An action group then produces new process description with elements that can be synthesized as "estimate; develop with increments; bi-weekly status review;". From process modeling point of view, this is a Process Enactment Description specification model.

After the implementation of this improved process in some pilot projects, three process descriptive modeling are performed. Each one produces a descriptive model of the new current process: (a) a Process Enacting Description model represented as "..., estimate; develop with increments; bi-weekly status review; ...", that answer the question "How we are doing"; (b) a Process Performance Indicator model represented as "About 80% projects are on time in a 20% range" that answer the question "What is going on"; and (c) a Process Capability Profile model represented as "most elements of extended 15504 ML2", that answer the question "Which best practices we are using". Then further sub-cycles are performed with more process modeling.

Finally, at the end of Institutionalize improvement phase another three process descriptive modeling are performed on the institutionalized improved process. Each one produces a descriptive model of this process: (a) a descriptive Process Enacting Description model represented as "..., estimate using Use Case Points; develop with increments; weekly status review; ..."; (b) a descriptive Process Performance Indicator model represented as "92% projects are on time in a 8% range"; and (c) a Process Capability Profile model represented as "extended 15504 ML2".

## 5 CMMI, SPICE and Modeling View of Process Improvement

ISO/IEC 15504 (SPICE) and CMMI are well established, relevant and representative of the current state of the art of SPI. From CMMI Framework, the SPI state of the art can be represented by IDEAL cycle for process improvement, SCAMPI method for process assessment [8] and CMMI-DEV model for development [3]. From ISO/IEC 15504 (SPICE) the SPI state of the art can be represented by the measurement framework for process capability, the measurement framework for organizational maturity, the requirements for performing an assessment (ISO/IEC 15504-2), the requirements for a Process Reference Model (PRM), the requirements for a Process Assessment Model (PAM), the steps of process improvement (ISO/IEC 15504-4), the exemplar PAM for software engineering (ISO/IEC 15504-5) and the exemplar organizational maturity model for software engineering (ISO/IEC 15504-7) [4].

As an evaluation of the claim that the modeling view of process and process improvement presented in the previous section, is an integrated view of current SPI, Table 1 and the next paragraphs provides an analyses on how CMMI and SPICE already cover these three types of process models and modeling. In Table 1, the column Elements indicates which CMMI and SPICE elements deals with each type of process model as specification and descriptive modeling. The column # indicates a degree of achievement of the full modeling with the set of elements. This degree is in the same four points scale defined by SPICE with similar meaning: N (Not achieved), P (Partially achieved), L (Largely achieved) and F (Fully achieved).

**Table 1.** Types of models/modeling and CMMI/SPICE models

| Types of models and modeling | | CMMI-DEV, IDEAL and SCAMPI | | ISO/IEC 15504 (SPICE) | |
|---|---|---|---|---|---|
| Model | Modeling | Elements | # | Elements | # |
| Process Capability Profile model | Specification modeling | IDEAL cycle; CMMI-DEV model | L | 15504-4 cycle; 15504-5;7 models | F |
| | Descriptive modeling | SCAMPI method; CMMI-DEV model | P | 15504-2 reqs. asses.; 15504-5;7 models | L |
| Process Enactment Description model | Specification modeling | OPD SP 1-4; IPM SP 1.1 | F | MAN.1.BP2; PIM.1.BP1,BP3 | F |
| | Descriptive modeling | PPQA SG 1 | L | SUP.1.BP3; SUP.5.BP3 | L |
| Process Performance Indicator model | Specification modeling | MA SG1 | F | MAN.6.BP3,BP4 | F |
| | Descriptive modeling | MA SG2 | L | MAN.6.BP5,BP6 | L |

Each CMMI-DEV profile, including maturity levels 2 to 5, is a Process Capability Profile type of model. During the first phases of a process improvement cycle, following, for example, the IDEAL model, a CMMI-DEV profile is used as a Process Capability Profile specification model for the future process. This profile however uses only a predefined set of the 22 process areas defined in CMMI-DEV. Only a process that is an implementation of these process areas can be specified. Any other relevant process cannot be specified. Therefore a CMMI-DEV profile cannot always be a model of the whole future process. So CMMI Largely (L) achieves Process Capability Profile specification modeling. An assessment using SCAMPI method uses a CMMI-DEV profile as reference. During the first phases of SCAMPI, a specific CMMI profile is defined. A SCAMPI assessment produces a Process Capability Profile descriptive model only for the process that implements that profile. Any relevant process that is not an implementation of that profile is not considered. So CMMI Partially (P) achieves Process Capability Profile descriptive modeling.

Each 15504-5 profile, and each 15504-7 organizational maturity levels from 1 to 5, is a Process Capability Profile type of model. Similar with CMMI, with a difference: It is possible to define any new process for 15504-5 and included it in a 15504-7 organizational maturity model. Therefore SPICE Fully (F) achieves Process

Capability Profile specification modeling. For assessment, however, the profile must be specified during the first phases of an assessment. The descriptive model is constrained by this profile. Therefore SPICE Largely (L) achieves Process Capability Profile descriptive modeling.

CMMI-DEV´s Organizational Process Definition (OPD) and Process and Product Quality Assurance (PPQA) process areas include modeling with Process Enactment Description type of model. CMMI defines "process description" as "a documented expression of a set of activities performed to achieve a given purpose". The term "Process Enactment Description model" is used in this article to mean CMMI process description. The word "enactment" is necessary because all types of process models are process descriptions. This specific description is to guide the enactment of a process. CMMI defines "process definition" as "the act of defining and describing a process" and its result as "process description". The term "Process Enactment Description modeling" is used in this article to mean CMMI process definition.

The purpose of OPD is "to establish and maintain a usable set of organizational process assets, work environment standards, and rules and guidelines for teams". OPD SG (Specific Goal) 1 states "a set of organizational process assets is established and maintained". OPD SP (Specific Practice) 1.1 states "establish and maintain the organization's set of standard processes". OPD SP 1.2 states "establish and maintain descriptions of lifecycle models approved for use in the organization". OPD SP 1.3 states "establish and maintain tailoring criteria and guidelines for the organization's set of standard processes". The purpose of IPM is "to establish and manage the project and the involvement of relevant stakeholders according to an integrated and defined process that is tailored from the organization's set of standard processes". IPM SG1 states "the project is conducted using a defined process tailored from the organization's set of standard processes". IPM SP 1.1 states "establish and maintain the project's defined process from project startup through the life of the project". So CMMI Fully (F) achieves Process Enactment Description specification modeling

The purpose of PPQA is "to provide staff and management with objective insight into processes and associated work products". PPQA SG 1 states "Adherence of the performed process and associated work products to applicable process descriptions, standards, and procedures is objectively evaluated". This evaluation is constrained by to applicable process descriptions, standards, and procedures, so the result can be a descriptive partial model of the process. Therefore CMMI Largely (L) achieves Process Enactment Description descriptive modeling.

ISO/IEC 15504-5´s Organizational alignment process (MAN.1), Process establishment process (PIM.1), Quality assurance process (SUP.1) and Audit process (SUP.5) include modeling with Process Enactment Description type of model. The purpose of the Organizational alignment process (MAN.1) is to enable the software processes needed by the organization to provided software products and services, to be consistent with its business goals". MAN.1.BP2 states "Define the process framework - Identify the processes that need to be performed in order to achieve the business goals". The purpose of the Process establishment process (PIM.1) is to "establish a suite of organizational processes for all life cycle processes as they apply to its business activities". PIM.1.BP1 states "Define process architecture - Define a standard set of processes, purpose of each process and interactions between them" and PIM.1.BP3 states "Define standard processes - Define and maintain a description of each standard process according to the needs to establish processes in the organization (NOTE: Effective, organization-wide establishment of standard processes may require that they are documented)". Therefore SPICE Fully (F) achieves Process Enactment Description specification modeling.

The purpose of Quality assurance process (SUP.1) is to provide assurance that work products and processes comply with predefined provisions and plans. SUP.1.BP3 states "assure the quality of project process activities and project work products". The purpose of Audit process (SUP.5) is to independently determine compliance of selected products and processes with the requirements, plans and agreement, as appropriate. SUP.5.BP3 states "audit for conformance against the requirements. Selected work products, services or processes are audited to determine their conformance with their requirements and planned arrangements. Non-conformances are recorded". This assurance and this audit are constrained by predefined provisions and plans, so the result can be a descriptive partial model of the process. Therefore SPICE Largely (L) achieves Process Enactment Description descriptive modeling.

CMMI-DEV´s Measurement and Analysis (MA) process area and SPICE´s Measurement process (MAN.6) include modeling with Process Performance Indicator type of model. MA SG 1 states "Analysis Activities Measurement objectives and activities are [establish, maintain and] aligned with identified information needs and objectives". Therefore CMMI Fully (F) achieves Process Performance Indicator specification modeling. MAN.6.BP3 states "Identify measurement information needs - Identify the measurement information needs of organizational and management processes" and MAN.6.BP4 states "Specify measures - Identify and develop an appropriate set of measures based on measurement information needs". Therefore SPICE Fully (F) achieves Process Performance Indicator specification modeling.

MA SG 2 SG 2 states "Measurement results, which address identified information needs and objectives, are provided". These measurements results are constrained by predefined measures, so the result can be a descriptive partial model of the process. Therefore CMMI Largely (L) achieves Process Performance Indicator description modeling. MAN.6.BP5 states "Collect and store measurement data - Identify, collect and store measurement data, including context information necessary to verify, understand, or evaluate the data" and MAN.6.BP6 states "Analyze measurement data - Analyze and interpret measurement data, and develop information products". These measurements results are constrained by predefined measures, so the result can be a descriptive partial model of the process. Therefore SPICE Largely (L) achieves Process Performance Indicator description modeling.

# 6 Towards a Modeling Driven Process Improvement

The analyses described in the previous section indicates that CMMI and SPICE already cover these three types of process models and modeling, with three limitations: a) CMMI allows the usage of only a predefine set of process areas for Process Capability Profile models; b) CMMI and SPICE constrain descriptive modeling by previously specified specification models; and c) CMMI and SPICE did not fully explore the integration of all three types of process models.

Current SPI usually considers Process Capability Profile models as reference for process improvement, but neither Process Enactment Description models nor Process Performance Indicator models. Current SPI usually considers Process Enactment Description models as models of the processes, but neither Process Capability Profile models nor Process Performance Indicator models. Best practices models from CMMI and SPICE, as, for example, CMMI and ISO/IEC 15504-5, include a process area for Process Performance Indicator modeling (measurement process area), but neither for Process Capability Profile modeling nor Process Enactment Description modeling. All three types of models as references for process improvement and as process models could be considered to improve process improvement. Process areas for all three types of modeling could be considered as well.

The example in Section 4 indicates that using this modeling view during a process improvement cycle improves the integrated understanding of what we want (using specification modeling), what is the current status (using descriptive modeling), what we can do (using a balance between specification and descriptive models) and what we are doing (using specification and descriptive modeling) for improvement during a cycle. However, in order to use the whole potential of this modeling view, an improved worldview is proposed: Modeling driven Process Improvement. The term driven is used in the sense of Model Driven Engineering (MDE).

Modeling driven Process Improvement is the worldview of PRO2PI Methodology. PRO2PI Methodology has been evolved from **Pro**cess Capability **Pro**file *to* drive **P**rocess **I**mprovement [6] [14] towards **Pro**cess *Modeling* **Pro**file *to* drive **P**rocess **I**mprovement, in order to explore the whole potentiality of Modeling driven Process Improvement. This evolution emerged when a method for tridimensional process assessment using modeling theory was developed and used following Potts´s industry-as-laboratory approach [15]. Modeling driven Process Improvement is a worldview in which process improvement is driven by process modeling. There is a co-evolution of process models and process, where dynamic specification or descriptive process models (Process Capability Profile models, Process Enactment Description models and Process Performance Indicator models) represent a (specified) future process or the actual current process. In a previous article [6] the expression "model driven" was used. The expression "modeling driven" emphasizes the process of modeling instead of the model itself. The SPI Manifesto states three values and principles. The proposed Modeling driven Process Improvement is consistent with the SPI Manifesto´s values and principles, especially with the principle "Use dynamic and adaptable models as needed" using modeling as the underlying integration theory.

# 7 Conclusion

This article introduces a modeling view of process and process improvement with three types of process models (Process Capability Profile, Process Enactment Description and Process Performance Indicator) and (knowledge working) process as a model of what knowledge workers do and think for a living, where the model is a set of interrelated (or interacting) activities which transform inputs into outputs, to achieve a given purpose". Drucker´s knowledge working process is proposed as a generalization of software, systems, services and other domains processes. This modeling view is corroborated by an example of its usage in a process improvement cycle and by an analysis on how two representative SPI frameworks (CMMI and ISO/IEC 15504) partially support this view. Using this modeling view as a basis, Modeling driven Process Improvement is proposed as an evolution of current Model-based Process Improvement.

## References

1. Humphrey, W.S.: Managing the Software Process. Addison-Wesley, Reading (1989)
2. Rout, T.P., El Emam, K., Fusani, M., Goldenson, D., Jung, H.-W.: SPICE in retrospect: Developing a standard for process assessment. J. Syst. Software (2007)
3. CMMI Product Team, CMMI® for Development, Version 1.3, Improving processes for developing better products and services. Technical Report, CMU/SEI-2010-TR-033, ESC-TR-2010-033, Software Engineering Process Management Program (November 2010)
4. The International Organization for Standardization and the International Electrotechnical Commission. ISO/IEC 15504, composed of seven parts (15504-1 to 15504-7) parts, under the general title Information technology — Process assessment (2004-2008)
5. Dorling, A.: Next Generation 15504 - the 33001 series of Standards – UPDATE, August 24 (2009), http://www.spiceusergroup.org
6. Salviano, C.F.: Model-Driven Process Capability Engineering for Knowledge Working Intensive Organization. In: Proc. of 8th Int. SPICE Conf., Nuremberg, Germany, pp. 1–9 (2008)
7. Pries-Heje, J., Johansen, J. (chief eds.): SPI Manifesto, eurospi. net, version A.1.2 (2010)
8. Potts, C.: Software-Engineering Research Revised. IEEE Sw. 10(5), 19–28 (1998)
9. Bézivin, J.: On The Unification Power of Models. In: Software and System Modeling (2005)
10. Favre, J.M.: Megamodeling and Etymology - A story of Words: from MED to MDE via MODEL in five millenniums, ADELE Team, LSR-IMAG, University of Grenoble, France (2004), http://www-adele.imag.fr/~jmfavre
11. Rothenberg, J.: AI, Simulation & Modeling. In: Widman, L.E., Loparo, K.A., Nielsen, N.R. (eds.) The nature of modeling, August 1989, pp. 75–92. John Wiley & Sons, Inc., Chichester (1989); (Reprinted as N–3027–DARPA, The RAND Corporation, November 1989)
12. Atkin, A.: Peirce's Theory of Signs. The Stanford Encyclopedia of Philosophy (Winter 2010),
http://plato.stanford.edu/archives/win2010/entries/peirce-semiotics/
13. Drucker, P.: Landmarks of Tomorrow - A Report on the New 'Post-Modern' World. Harper & Row, New York (1959)
14. Salviano, C.F.: A Multi-Model Process Improvement Methodology Driven by Capability Profiles. In: Proc. of IEEE COMPSAC, Seattle, USA, pp. 636–637 (2009), doi:10.1109/COMPSAC.2009.94
15. Salviano, C.F., Martinez, M.R.M., Banhesse, E.L., Enelize, A., Zoucas, A., Thiry, M.: A Method for Tridimensional Process Assessment Using Modelling Theory. In: Proc. of IEEE Seventh QUATIC, Porto, Portugal, pp. 430–435 (2010), doi:10.1109/QUATIC.2010.95
16. Card, D.N.: Research Directions in Software Process Improvement. In: Proc. 28th IEEE Int. Comp. Sw. and App. Conf., Hong Kong, China, September 27-30, pp. 238–239 (2004)

# An Approach to Evaluating Software Process Adaptation

Paul Clarke[1] and Rory V. O'Connor [2,3]

[1] Lero Graduate School in Software Engineering, Dublin City University, Ireland
pclarke@computing.dcu.ie
[2] Dublin City University, Ireland
[3] Lero, the Irish Software Engineering Research Centre
roconnor@computing.dcu.ie

**Abstract.** Process maturity reference frameworks such as ISO/IEC 15504 and the Capability Maturity Model Integrated (CMMI) seek to assist software process improvement (SPI) efforts by prescribing a roadmap for improving the capability of the development process. However, such frameworks are not widely adopted in the practice [1], [2], especially in smaller software development organisations where the development process is often modified based on business events [3]. Such modification of the development process represents an attempt to harmonise the process with the changing needs of the business, which is a dynamic capability. Dynamic capabilities refer to the ability of businesses to adapt to changing circumstances and according to the *evolutionary* theory of the firm [4], organisations that possess greater dynamic capability are more successful. This paper introduces *dynamic SPI capability* - the ability to adapt the software process relative to changing situational circumstances – as a method for evaluating software process adaptation.

**Keywords:** SPI, Process Adaptation, Situational Factors, Dynamic SPI Capability.

## 1 Introduction

The past two decades have witnessed significant growth in the software development business and in parallel there has been a sustained investment in research into the process of software development. One of the principal developments in the software process domain has been the emergence of process maturity frameworks, with ISO/IEC 15504 [5] and CMMI [6] considered the most dominant [7-9]. Although process maturity reference frameworks are sometimes criticised for being cumbersome and costly [10], ISO/IEC 15504 [5] and CMMI [6] have been shown to deliver significant benefits for software development [11-16].

While the benefits of process maturity frameworks and other software development models have been demonstrated, they are not widely adopted in practice [1-3], [17], [18]. Furthermore, some research suggests that temporal contextual factors are critical in identifying the most appropriate process [10], [19], [20], especially in small to medium sized enterprises (SMEs) [21]. It is therefore not surprising to discover that the software development process has been reported as being volatile and that process improvements are often initiated in response to business events [3]. The ability of a business to learn from business events and to improve business processes to address changing needs is what economists describe as a dynamic capability – and according to the *evolutionary* theory of the firm [4], organisations that have greater dynamic capability are more likely to be successful. Therefore, a dynamic capability to improve the software development process with respect to changing circumstances, a characteristic which we have defined as *dynamic SPI capability*, is beneficial for business success in software development organisations.

Given that the dynamic SPI capability of an organisation is important for business success, it would be advantageous for organisations to be able to observe their dynamic SPI capability. However, to date SPI research efforts have been concerned largely with process capability rather than with dynamic process capability and as a result there is no existing mechanism for observing the dynamic SPI capability of a software development setting. Therefore, this paper identifies the key considerations when attempting to determine the dynamic SPI capability in a software development setting and proposes an approach to unifying these considerations so as to bring visibility to this important capability.

This paper is structured as follows: Section two provides additional background information on the relevance of dynamic capability to software development organisations. Sections three and four outline approaches to making determinations in relation to the two components of dynamic SPI capability: extent of SPI activity and extent of situational change. Section five identifies related future work that the authors intend to carry out and finally, section six presents a discussion and conclusion.

## 2 Dynamic Capability

In the field of economics, the evolutionary theory of the firm [4] is concerned with the concept of dynamic capability. Dynamic capability relates to the ability of an organisation to continually transform the business routines in response to changing environments and new understandings, and the evolutionary theory of the firm suggests that this ability gives rise to the dynamism that will ultimately propel the organisation to success [22]. The firm, therefore, is promoted as "a locus where competencies are continually built, managed, combined, transformed, tested and selected", where the vital consideration relates to how "new knowledge [is] materialised in new competencies", and where "a lock-in to inefficient routines" is perceived as a major threat to a company's prospects [23]. Consequently, a dynamic capability to transform routines is considered to provide a basis for competitive advantage [23], a point that has already been observed in relation to the software development routines by Poulin [24], who suggests that with respect to software process capability, establishing an organisation's ability to optimise the development process may provide a better approach than traditional audits. Therefore, rather than examining process capability and prescribing an improvement path, an alternative view suggests that one should focus on maximizing the capability to transform the process, and that this transformational capability will automatically render an improved process.

The initial stage on a process maturity roadmap generally represents a state of low process capability, with subsequent stages gradually enhancing the process implementation, finally culminating with the process optimisation stage, wherein the software development process is continually being optimised in order to best address the software development needs of the organisation. Therefore, in a sense, existing process maturity frameworks do consider dynamic SPI capability – but only at the highest level of maturity. However, where process maturity reference frameworks are adopted, the highest maturity level (optimising) is not often achieved [25]. Furthermore, the evolutionary theory of the firm establishes that it is dynamic capability, rather than process maturity, that is of paramount importance for optimum process adaptation. Perhaps this raises the case for process optimisation to be brought forward as a consideration in process maturity frameworks, into a position where process optimisation is an inherent consideration for every process decision at every level in a process maturity framework. This consideration is, however, one for the committees responsible for maintaining the various process maturity frameworks – the focal point of this paper is to outline the components of dynamic SPI capability along with some mechanisms for determining the dynamic SPI capability of a software development setting.

In order to determine the dynamic SPI capability of a software development setting, two principal components must be established: (1) the extent of SPI activity and (2) the degree of situational change. The following sections discuss these two key components.

## 3 Extent of SPI Activity

In order to examine the dynamic capability with respect to the software development process, it is first necessary to determine the extent of SPI activity in an organisation. For the purpose of this paper, SPI activity is defined as *"the set of SPI actions implemented by an organisation, which is manifested as a series of modifications to the software development process"*. A review of the literature in the process assessment and auditing domain reveals that there is no dedicated approach to expressly examining the amount of software development process change that has occurred over a period of time. By *expressly,* we mean (1) in a single engagement and (2) collecting only data that is related to the extent of process change (not process capability). Since no express method for determining the amount of SPI activity pre-existed, it was necessary to develop a dedicated SPI activity survey instrument. Such a survey instrument would need to be systematically derived from a comprehensive and recognized software development process reference framework. ISO/IEC 12207 [26], which has been developed and maintained by international consensus, offers an ideal reference framework for the construction of an SPI activity survey instrument.

The ISO/IEC 12207 [26] based SPI activity survey instrument developed and presented below can be used to examine the extent of SPI activity. This instrument is different from traditional process assessments in that it directly and explicitly examines the extent of SPI actions and is not concerned with making process capability determinations. With the software development process constituting an important and complex component of the overall business process for software developing organisations, and acknowledging the importance of dynamic process capability as encapsulated in the evolutionary theory of the firm [4], software development and quality management practitioners, as well as auditing agents, can apply the SPI activity survey instrument in order to directly determine the extent to which the software development process is being evolved. Researchers can also use the SPI activity survey instrument, and the authors of this paper are presently applying the approach as part of a broader research project that is examining the influence of SPI on the evolution of small to medium sized (SME) software development companies.

It is possible to utilise the process assessment vehicles associated with process maturity reference frameworks in order to determine the amount of SPI activity. This would involve conducting two process assessments on two different dates, and thereafter performing a finite difference analysis on the assessment results. However, this twin assessment approach has a number of drawbacks. Firstly, it requires two engagements with the software development organisation, which is time consuming and which can be difficult to orchestrate from a practical researching perspective. Secondly, process assessments, such as those in ISO/IEC 15504 [5] and CMMI [6] collect data and generate a maturity ratings rather than just investigating the amount of SPI activity, and therefore represent a somewhat inefficient tool for evaluating SPI activity. Thirdly, adopting an ISO/IEC 15504 [5] or CMMI [6] process assessment vehicle to determine SPI activity might diminish the capacity to secure candidate participants in the SME sector, since prescribed process maturity reference frameworks have themselves already met with resistance to implementation in SMEs. For these three reasons, traditional process assessments would represent an inefficient use of resources for the purpose of making express SPI activity determinations.

Taking these drawbacks into account, and owing to the apparent absence of any established dedicated resource for determining the amount of SPI activity, we developed a new method for evaluating SPI activity, a method based around the application of a dedicated SPI activity survey instrument.

## 3.1 Evaluating SPI Activity Using a Dedicated Survey Instrument

In the case of ISO/IEC 15504 [5], the ISO/IEC 12207 [26] process listing is used as the underlying process reference list. ISO/IEC 12207 [26] is an internationally developed and maintained listing for software processes and therefore represents a useful reference point when examining software processes in any setting.

It is the premise of this paper that in order to evaluate the amount of SPI activity in an organisation, ISO/IEC 12207 [26] can be used as a comprehensive point of reference. However, the creation of a survey instrument based on ISO/IEC 12207 [26] needs to be structured and systematic, and this paper presents an approach suited to converting an international standard into a survey instrument, followed by an explanation of how the method was applied in the case of transforming ISO/IEC 12207 [26] into an appropriate survey instrument for evaluating the extent of SPI activity in an organisation.

### 3.1.1 Method for Converting an International Standard to a Survey Instrument

Many international standards consist of verbose text that seeks to accurately and completely describe an item of technical matter. However, such comprehensive text-based descriptions are not easily fashioned into survey instruments, especially when practical considerations, such as the time required to conduct the survey, are taken into consideration. Therefore, this paper outlines a technique for resolving verbose text-based international standards back to comprehensive, yet practical, survey instruments. An overview of this technique is presented in figure 1.

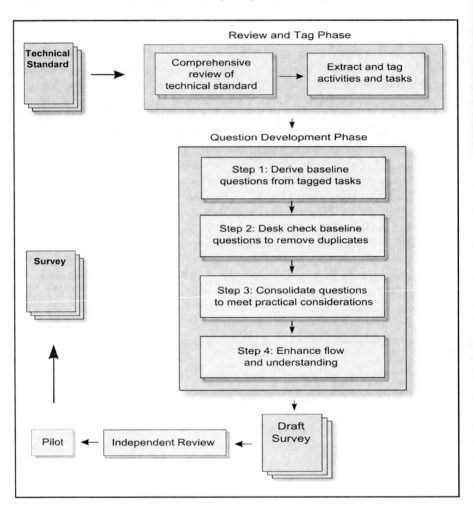

**Fig. 1.** Survey Instrument Development Technique

The initial phase, the *Review and Tag* phase, involves reviewing the international standard, so as to develop a thorough understanding of all the material comprising the standard. Thereafter, the various components of the international standard are tagged – in order to identify the key activities. This requires that close attention is paid to all actions in the international standard, ensuring that no important detail is overlooked.

Following the tagging exercise, the *Question Development* phase is undertaken. This is a four-step activity that involves transforming the tagged details, as output from the initial phase, into a representative, accurate, comprehensive and readable survey instrument. Notes that explain any modifications, along with rationale for changes, must be maintained at each step in the question development phase – this allows for later examination of the survey construction exercise, including the possibility of auditing the artefacts so as to verify that appropriate decisions have been taken throughout the survey construction activity. Such artefacts can thereafter be published along with the survey findings if required.

The first step of the question development phase involves using the tagged details in order to derive a baseline set of questions. This results in a baseline suite of questions that preserve all of the essential details that are present in the international standard itself. In the second step of the question development phase, the baseline suite of questions is desk-checked so that any duplications or areas of overlap are resolved. This is necessary in order to efface cross-references that can exist in international standards.

The third step of the question development phase consolidates the list of questions with respect to practical considerations. The target survey duration is among the practical considerations, and the survey constructor must judge the appropriate type and number of questions for the survey. The consolidation of questions also requires a considerable deal of judgement, coupled with expertise, on the part of the survey constructor, but should nonetheless seek to preserve the original makeup and structure of the international standard, retaining all major components such that the resulting survey is clearly identifiable as a derivative of the original standard. Having consolidated the questions in an appropriate fashion, the fourth and final step of the question development phase involves reviewing the survey so as to enhance the clarity of individual questions and to optimise the flow of the survey so as to best achieve the survey objectives.

Having completed the question development phase, the survey constructor presents a draft version of the survey instrument to software process and process standards domain experts so as to elicit independent feedback on the content, accuracy, and likely effectiveness of the interview in obtaining the required information. Following completion of the independent review, the survey instrument should be revised so as incorporate the feedback from the expert reviewer. Once again, a copy of the changes applied should be maintained so as to allow for later examination of the technique.

### 3.1.2 Application of Conversion Method to ISO/IEC 12207

The technique identified in figure 1 was applied to ISO/IEC 12207 [26] and a systematically-derived draft SPI activity survey instrument was produced. This draft survey instrument was submitted to key ISO/IEC 12207 [26] editorial committee members for review, after which a final rendering of the SPI activity survey instrument was produced. Further details of the survey instrument creation can be found in [27]. In producing the SPI activity survey instrument, a detailed review of ISO/IEC 12207 [26] was carried out. Following this review, a diagrammatical overview of the activities contained in ISO/IEC 12207 [26] was produced – this overview is reproduced here in figure 2.

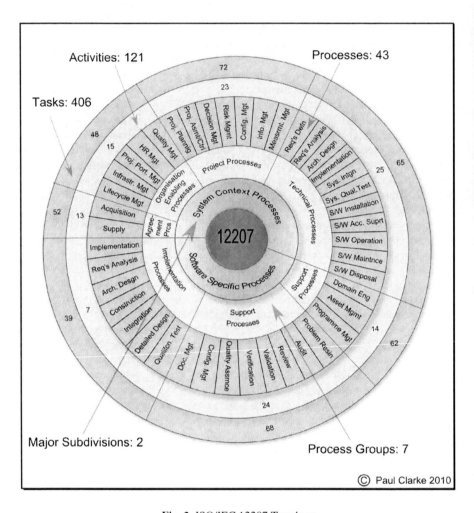

Fig. 2. ISO/IEC 12207 Topology

This section has outlined the approach adopted in constructing an express SPI activity survey instrument from ISO/ISE-12207 [26]. However, in order to determine the dynamic SPI capability of a software development setting, it is also necessary to determine the extent of situational change that has occurred.

## 4 Extent of Situational Change

In order to examine the extent of situational change in a software development setting, it is necessary to utilise a comprehensive reference framework of the factors that affect the software development process – rather than being concerned with changes to the general environment, in examining dynamic SPI capability, we are only interested in changes to the environment that are potential triggers for SPI. However, a literature review of the domain confirms that no single general and comprehensive reference framework of the situational factors that affect the software development process presently exists. Therefore, the authors have systematically developed such a reference framework of the factors that affect the software development process [28]. As with the development of the express SPI activity survey instrument outlined earlier, a systematic approach was adopted in the development of the reference framework of the situational factors that affect the software development process. This involved a synthesis of twenty-two of the most influential works from seven distinct domains: software development models and standards, risk factors for software development, software development cost estimation, software development environmental factors, software process tailoring, degree of required software process agility, and the software engineering book of knowledge [29]. The approach to developing the reference framework of the factors affecting the software process is outlined in the figure 3.

The framework construction exercise utilised numerous data sources, including Boehm's Cost Constructive Model (CoCoMo) [30], Putnam's SLIM model [31], Albrecht's Function Point Analysis (FPA) [32], CMMI [6], and ISO/IEC 12207 [26]. Following the *Review and Tag* phase, a baseline of three hundred and ninety-seven factors affecting the software development process was identified. Since the data sources are thematically related, this initial baseline contained some duplication – both literal and conceptual – and therefore it was necessary to distil a consolidated reference framework. This distillation exercise required a concentrated data analysis effort and therefore, data analysis techniques from Grounded Theory [33] were applied so as to ensure that a rigorous and systematic approach was adopted. Borrowing the *constant comparison* [34] and *memoing* [35] data analysis techniques from Grounded Theory, the baseline set of factors was systematically consolidated into a comprehensive set of forty-four individual situational factors that affect the software development process. These factors are classified under eight categories and have a total of one hundred and fifty-seven sub-factors, as outlined in figure 4.

### 4.1 Evaluating Situational Change Using a Dedicated Survey Instrument

The situational factors affecting the software development process indicated in figure 4 are used as a reference framework for the development of a comprehensive survey instrument that will determine the extent of situational change that has occurred over a period of time. In keeping with the approach adopted in the earlier SPI activity survey instrument, again a systematic technique is adopted. The *Question Development*, *Independent Review* and *Pilot* phases identified in figure 1 are re-used – this time to develop a survey instrument from the reference framework of the factors affecting the software development process. The resulting survey instrument will facilitate a determination of the amount of situational change that has occurred by examining changes to the factors affecting the software development process over a period of time.

In section 3, we introduced an express approach to determining the amount of SPI activity in a software development setting over a period of time. This section has identified an approach to determining the extent of situational change in a software development setting. In order to determine the dynamic SPI capability in a software development setting, we must unify these two components.

## 5 Future Work - Dynamic SPI Capability

As outlined earlier, dynamic SPI capability relates to the ability to adapt the software development process in tune with changing situational circumstances. Furthermore, two key components are required in order to determine the dynamic SPI capability in a software development setting: (1) the extent of SPI activity, and (2) the extent of situational change. Earlier sections have identified approaches to determining both of these components. The next stage in our research involves the identification of an appropriate method for integrating these two components into a form that enables the visualisation of the dynamic SPI capability – this is a work in progress but at present there are two candidate approaches.

Firstly, the two distinct components could be unified into a single key performance indicator (KPI). This would involve expressing the one component in a ratio form relative to the other component. There are a number of benefits to having the dynamic SPI capability accessible in a single KPI numeric form; for example, it would be possible to easily compare the dynamic SPI capability over time in different settings and it would be permissible to offer a guidance scale of performance in relation to SPI. However, there are some challenges to integrating these two components into a single KPI: for example, the two components are quite different in nature and therefore, it may not be useful to attempt to integrate them into a single numeric form.

Secondly, the two distinct components could be visualized in a four-quadrant type graph, where the first dimension would depict the extent of the SPI activity and the second dimension would capture the extent of situational change. This particular approach would overcome the noted challenge in relation to unifying the two components into a single numeric form – however, it also has some drawbacks; for instance, it would be more difficult to compare the dynamic SPI capability in two different settings.

Over the coming months, the authors will be actively examining the options for unifying the two components of dynamic SPI capability that have been outlined in this section.

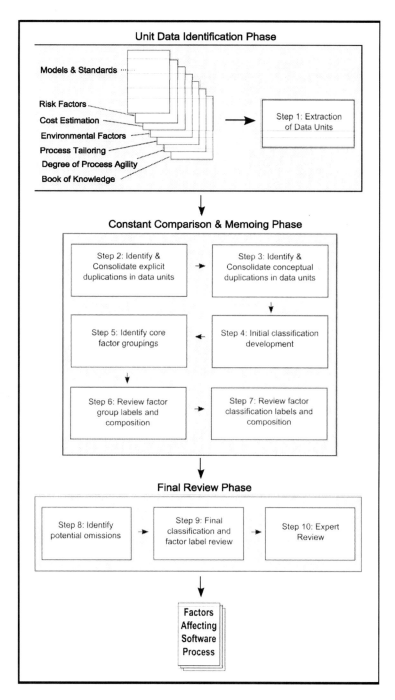

**Fig. 3.** Situational Factors Reference Framework Development Technique

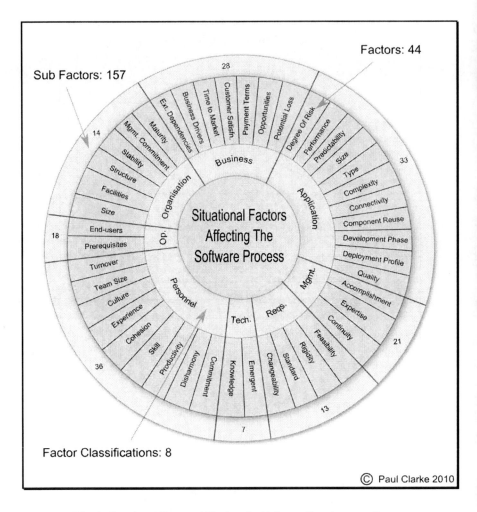

**Fig. 4.** Situational Factors Affecting the Software Development Process

## 6 Discussion and Conclusion

The quality of the software development process directly affects the quality of the software product, and since the technology, business environment and company circumstances are subject to continual change, there is an ongoing requirement for SPI. Existing approaches to SPI, such as ISO/IEC 15504 [5] and CMMI [6] assess the capability of processes in an organisation. These process maturity reference frameworks prescribe a phased process maturity roadmap, with the earlier stages characterised by minimum process implementation and the later stages gradually improving the process maturity, with the final stage being dedicated to continuous process optimisation.

The concept of process optimisation is related to the evolutionary theory of the firm [4], which suggests that the dynamic capability of an organisation to modify its business processes is an important driver for business success. If it is the case that dynamic capability is central to the formula for business success, then software development organisations would benefit from being dynamically capable with respect to the software development process. Process maturity frameworks such as ISO/IEC 15504 [5] and CMMI [6] do acknowledge process optimisation as an important attribute, but it is only evident at the most mature stage. Therefore, organisations that adopt such process maturity references frameworks, and who do not progress to the most mature stage, may fail to realise the benefits of dynamic capability as described by the evolutionary theory of the firm [4].

If dynamic capability is important, and we suggest that it is, then there should be a method for examining the dynamic process capability in an organisation. For software development organisations, this includes the ability to examine dynamic capability with respect to the software development process, or to use the term introduced in this paper: *dynamic SPI capability*. In order to observe dynamic SPI capability, it is necessary to determine the amount of SPI that has taken place in a software development environment setting over a period of time, and to concurrently examine the changes in the situational context that are important considerations for the software development process. This paper outlines two mechanisms: one for making express determinations in relation to the amount of SPI activity that has occurred and a second mechanism that enables the determination of the amount of change that has occurred in the situational context. Together, these two mechanisms can be combined to make an overall determination in relation to the dynamic SPI capability of a software development setting over a period of time.

It is not the intention of the approach outlined in this paper to devalue the very significant benefits and contributions that are already available in capability maturity frameworks such as ISO/IEC 15504 [5] and CMMI [6]. However, as a research community we must accept that despite being in existence for a long number of years, these approaches are not widely adopted across the broader spectrum of the software development industry. Equally, where process maturity frameworks are implemented, they are often tailored to individual settings and software development settings themselves are subject to changing circumstances on a regular basis. Consequently, an important characteristic of a software development process is its ability to continually adapt to meet the needs of the changing environment – and one could argue that the amount of process adaptation required in a setting is a function of the degree of situational change in that setting. By developing an approach to examining situational change, and through the application of the dynamic capability concept to the software development process, this paper has presented a systematically-derived and novel approach to evaluating software process adaptation in a software development setting.

**Acknowledgments.** This work is supported, in part, by Science Foundation Ireland grant 03/CE2/I303_1 to Lero, the Irish Software Engineering Research Centre (www.lero.ie).

## References

1. McConnell, S.: Closing the Gap. IEEE Software 19(1), 3–5 (2002)
2. Staples, M., Niazi, M., Jeffery, R., Abrahams, A., Byatt, P., Murphy, R.: An Exploratory Study of Why Organizations do Not Adopt CMMI. Journal of Systems and Software 80(6), 883–895 (2007)
3. Coleman, G., O'Connor, R.: Investigating Software Process in Practice: A Grounded Theory Perspective. Journal of Systems and Software 81(5), 772–784 (2008)
4. Nelson, R.R., Winter, S.: An evolutionary theory of economic change. The Balknap Press of Harvard University Press, Cambridge (1982)
5. ISO/IEC: 15504-1 information technology - process assessment - part 1: Concepts and vocabulary. ISO / IEC, Geneva, Switzerland (2004)
6. CMMI Product Team. CMMI for development, version 1.2. Software Engineering Institute, CMU/SEI-2006-TR-008. Pittsburgh, PA, USA (2006)
7. Salviano, C.F., Figueiredo, A.: Unified Basic Concepts for Process Capability Models. In: Proceedings of The Twentieth International Conference on Software Engineering and Knowledge Engineering (SEKE 2008), pp. 173–178 (2008)
8. McBride, T., Henderson-Sellers, B., Zowghi, D.: Project Management Capability Levels: an empirical study. In: Proceedings of the 11th Asia-Pacific Software Engineering Conference, pp. 56–63 (2004)
9. Haapio, T.: A Framework for Improving Effort Management in Software Projects. Software Process: Improvement and Practice 12, 549–558 (2007)
10. Fayad, M.E., Laitnen, M.: Process Assessment Considered Wasteful. Communications of the ACM 40(11), 125–128 (1997)
11. Herbsleb, J., Carleton, A., Rozum, J., Siegel, J., Zubrow, D.: Benefits of CMM-Based Software Process Improvement: Initial Results. Software Engineering Institute, Carnegie Mellon University, Pittsburgh, Pennsylvania, USA (1994)
12. Herbsleb, J., Goldenson, D.: A systematic survey of CMM experience and results. In: Proceedings of the 18th International Conference on Software Engineering (ICSE 1996), pp. 323–330. IEEE Computer Society Press, Los Alamitos (1996)
13. Lawlis, P., Flowe, R., Thordahl, J.: A Correlational Study of the CMM and Software Development Performance. Crosstalk, The Journal of Defense Software Engineering 8(9), 21–25 (1995)
14. Gibson, D., Goldenson, D., Kost, K.: Performance results of CMMI-Based Process Improvement. Software Engineering Institute, Carnegie Mellon University, CMU/SEI-2006-TR-004, Pittsburgh, Pennsylvania, USA (2006)
15. Wegelius, H., Johansson, M.: Practical Experiences on Using SPICE for SPI in an Insurance Company. In: Abrahamsson, P., Baddoo, N., Margaria, T., Messnarz, R. (eds.) EuroSPI 2007. LNCS, vol. 4764, p. 2. Springer, Heidelberg (2007)
16. El Emam, K., Birk, A.: Validating the ISO/IEC 15504 Measures of Software Development Process Capability. Journal of Systems and Software 51(2), 119–149 (2000)
17. McAdam, R., Fulton, F.: The Impact of the ISO 9000:2000 Quality Standards in Small Software Firms. Managing Service Quality 12(5), 336–345 (2002)
18. Ludewig, J.: Software engineering in the years 2000 minus and plus ten. In: Wilhelm, R. (ed.) Informatics: 10 Years Back, 10 Years Ahead. LNCS, vol. 2000, p. 102. Springer, Heidelberg (2001)
19. Benediktsson, O., Dalcher, D., Thorbergsson, H.: Comparison of Software Development Life Cycles: A Multiproject Experiment. IEE Proceedings - Software 153(3), 87–101 (2006)

20. MacCormack, A., Verganti, R.: Managing the Sources of Uncertainty: Matching Process and Context in Software Development. Journal of Product Innovation Management 20(3), 217–232 (2003)
21. Kautz, K.: Software Process Improvement in very Small Enterprises: Does it Pay Off? Software Process: Improvement and Practice 4(4), 209–226 (1998)
22. Chandler, A.D.: Organizational Capabilities and the Economic History of the Industrial Enterprise. The Journal of Economic Perspectives 6(3), 79–100 (1992)
23. Cohendet, P., Kern, F., Mehmanpazir, B., Munier, F.: Knowledge Coordination, Competence Creation and Integrated Networks in Globalised Firms. Cambridge Journal of Economics 23(2), 225–241 (1999)
24. Poulin, L.A.: Achieving the Right Balance between Process Maturity and Performance. IEEE Canadian Review 56(-), 23–26 (2007)
25. Davenport, T.H.: The Coming Commoditization of Processes. Harvard Business Review, 100–108 (June 2005)
26. ISO/IEC. Amendment to ISO/IEC 12207-2008 - systems and software engineering – software life cycle processes. ISO, Geneva, Switzerland (2008)
27. Clarke, P., O'Connor, R.: Harnessing ISO/IEC 12207 to examine the extent of SPI activity in an organisation. In: Riel, A., et al. (eds.) EuroSPI 2010. CCIS, vol. 99, pp. 25–36. Springer, Heidelberg (2010)
28. Clarke, P., O'Connor, R.V.: The Situational Factors that Affect the Software Development Process. IEEE Transactions on Software Engineering (Submitted)
29. IEEE. Guide to the software engineering book of knowledge (SWEBOK). IEEE Computer Society, Los Alamitos (2004)
30. Boehm, B., Clark, B., Horowitz, E., et al.: Software cost estimation with cocomo II. Prentice-Hall PTR, Upper Saddle River (2000)
31. Putnam, L.: A General Empirical Solution to the Macro Software Sizing and Estimating Problem. IEEE Transactions on Software Engineering 4(4), 345–361 (1978)
32. Albrecht, A.J.: Measuring application development productivity. In: Proceedings of the IBM Applications Development Symposium, p. 83. GUIDE International and SHARE, Inc., IBM Corporation (1979)
33. Glaser, B., Strauss, A.: The discovery of grounded theory: Strategies for qualitative research. Aldine de Gruyter, Hawthorne (1976)
34. Corbin, J., Strauss, A.: Basics of qualitative research. Sage Publications Limited, Thousand Oaks (2008)
35. Bryant, A., Charmaz, K.: The SAGE handbook of grounded theory. Sage, Thousand Oaks (2007)

# Evaluation of Software Process Assessment Methods – Case Study

Mohammad Zarour[1], Alain Abran[2], and Jean-Marc Desharnais[3]

[1] Petra University
Department of Software Engineering
P.O. Box 961343
Amman 11196 Jordan
mzarour@uop.edu.jo
[2] École de Technologie Supérieure,
Department of Software and IT Engineering
1100 Notre-Dame Ouest, Montréal,
Québec H3C 1K3, Canada
Alain.abran@etsmtl.ca
[3] Computer Engineering
Bogazici University
34342 Bebek, Istanbul, Turkey
desharnaisjm@gmail.com

**Abstract.** Evaluation of competing software process assessment (SPA) methods is an important issue for software process improvement initiatives. Although SPA methods designers may claim successful design and implementations of their SPA methods, no evaluation of these claims, based on a set of evaluation criteria, has yet been documented. In addition, independent evaluation of the SPA methods currently available would also help the designers of these methods. This paper presents as a case study the results of applying the proposed evaluation criteria to the MARES SPA method.

**Keywords:** Software, Process, Assessment, Improvement, Evaluation.

## 1 Introduction

The technical literature documents a number of case studies discussing the implementation of software process assessment (SPA) methods, including lessons learned, success factors, requirements and observations. Although designers and users (assessors) of these SPA methods claim successful designs and implementations of these methods, there is a scarcity of documented independent verification and validation of these claims.

In the software engineering field, evaluations are often developed and performed without taking into account lessons learned from other software and non-software disciplines [1]. A well designed evaluation should be based on some theoretical basis, such asthe evaluation theory [2, 3]. The study of evaluation theories and methods [2-4] can help elaborating more complete and systematic evaluation methods for their application in the diverse software engineering areas [1].

Evaluation criteria have been proposed in [5] to help in evaluating software process assessment (SPA) methods, as well as in designing such SPA methods; [5] also presents a number of practices to develop a successful SPA method. The evaluation criteria can be used to evaluate any assessment method, and in particular those for Small Medium Enterprises (SME) as well as Very Small Entities (VSE) (those entities with less than 25 employees). This is illustrated in this paper with a case study of the evaluation of the MARES SPA method. The paper is organized as follows: Section 2 presents the main concepts of the MARES assessment method; Section 3 presents the evaluation procedure; Section 4 presents the evaluation results and Section 5 presents theconclusion and future work.

## 2 The MARES Assessment Method

The MARES (Methodology for Software Process Assessment in Small Software Companies)methodology has been built by researchers from UNIVALI University and the CenPRA research center in Brazil[6, 7].MARES is designed to support process improvement in the context of small software organizations considering their specific characteristics and limitations; this MARES model is built in conformity to ISO 15504 [8]. MARES method integrates a context-process model in order to support the selection of relevant processes and a process-risk model to support the identification of potential risks and improvement suggestions. The MARES method is discussed in detail in [6, 7, 9-13]. The MARES assessment method is divided into five main parts:

1. Planning:
   In this phase, the assessment is organized and planned. At the end of this phase, the resulting assessment plan is revised and documented.
2. Contextualization:
   In this phase, the organization is characterized in order to understand its goals, products and its software process. Questionnaires and interviews are used as a means to collect data.
3. Execution:
   The selected processes are assessed in detail.
4. Monitoring and control:
   All activities during the assessment are monitored and controlled. Corrective actions are initiated, if necessary, and the plan is updated accordingly.
5. Post-mortem:
   Once the assessment is finished, a brief post-mortem session is held among the assessors to discuss the performance of the assessment.

## 3 Evaluation Procedure

The evaluation of the MARES SPA method has been conducted in five main phases:

1. The evaluation criteria, see Appendix-A, have been used to construct a questionnaire as an evaluation tool, where each criterion is represented by a question having one of three possible answers: fully adequate, partially adequate and non adequate.

2. Answering the MARES questionnaire. One of the authors and main researchers in the design and deployment of the MARES assessment method was contacted and asked to answer the questionnaire, producing an PAM assessment report.
3. Documents analysis: various documents, presenting and discussing the MARES assessment method,were reviewed- see [6, 7, 9-13].
4. Analysis of the strengths and weaknesses of the assessed method and preparation of the assessment report.
5. Collection of feedback from the MARES' author about the findings presented in step 4 in order to validate the assessment.
6. Evaluation report submission: The final results of the assessment are documented.

## 4 Evaluation Results

The summary of the evaluation conducted based on the evaluation procedure presented in Section 3is shown in Table 1.

The evaluation results have been translated into percentages. For each question, a weight is assigned: 1 for fully adequate rating, 0.5 for partially adequate and 0 for non-adequate. The strengths points are achieved by summing up the number of fully adequate practices and the sum of the partially adequate practices. Similarly, the weakness points are achieved by summing up the number of non-adequate practices and the sum of the partially adequate practices. The percentages are then calculated for the strength points and weakness points.

As observed from Table 1, the MARES method fulfilled totally the best practices related to the 'procedure' class and fulfils mostly the best practices related to the 'users' class with around 83% of the practices for this class, while it did not fulfill any of the practices related to the 'supportive tool' since such a tool is not yet provided .

### 4.1 Strengths

The evaluation results based on the evaluation criteria in [5], see Appendix-A, show that the MARES method has several strengths which can be summarised as follows:

1. **SPA method:**
   a. Data gathering technique: MARES method collects the data through scheduled interviews with the organization participants.
   b. Flexible and customizable method that allows adding new processes to be assessed based on the organization's needs
   c. Coverage to a process reference model. The MARES process dimension has been developed based on ISO 15504-5 processes which are based on the ISO 12207. Due to the characteristics of small organizations, some processes have been disregarded as being irrelevant in most cases. If any of these disregarded processes turn out to be important, they are re-integrated based on the ISO 15504-5 as discussed in [7].
   d. Identification of strengths, weaknesses, risks and improvement opportunities.
   e. Suggest improvement action plan to start an improvement process.

**Table 1.** MARES Method Evaluation Results

| SPA Best Practices Categories | Total practices | Strengths Points | Strengths % | Weaknesses Points | Weaknesses % | % strengths contribution |
|---|---|---|---|---|---|---|
| Method (MBP) | 13 | 10.0 | 76.9% | 3 | 23.1% | 26.3% |
| Supportive tool (SBP) | 6 | 0.0 | 0.0% | 6 | 100.0% | 0.0% |
| Procedure (PBP) | 5 | 5.0 | 100.0% | 0 | 0.0% | 13.2% |
| Document. (DBP) | 8 | 6.5 | 81.3% | 1.5 | 18.8% | 17.1% |
| User (UBP) | 6 | 5 | 83.3% | 1 | 16.7% | 13.2% |
| **Total** | **38** | **26.5** | **69.7%** | **11.5** | **30.3%** | **69.7%** |

| | |
|---|---|
| Total practices | As shown in appendix A for each category |
| Strengths points | The count of practices scored as fully (1 point) or partially (0.5 point) in the evaluation |
| Strength % | Strengths points / total practices of the category |
| Weaknesses points | The count of practices scored as inadequate (0 points) in the evaluation |
| Weakness % | Weakness points / total practices of the category |
| % Strength contribution | Strengths points of each category / total number of practices (38 practices) |

    f. The method is publicly available including the process description and artefacts' templates. The method can be used in on-site/self assessment, yet requires an experienced assessor, who must be available on-site.

    g. Comply with a comprehensive assessment method: MARES is compliant with the assessment requirements as stated in ISO 15504. The MARES method assesses subsets of the processes that are relevant to the organization's needs.

    h. Simple and well-structured method having no more than 150 questions in its questionnaire.

2. **SPA Supportive Tools:**
   The MARES method did not fulfill any of the practices related to the supportive tool since no tool is available yet.

3. **SPA procedure:**
   a. Prepare the participant in the assessment. The MARES method, ISO 15504, the developed assessment plan and schedule are presented briefly to all assessment participants at the beginning of the assessment.
   b. The method includes building trust and confidence relationships between assessors and organization participants through the assessment briefing at the beginning as well as a confidentiality form to be signed by all assessors and the sponsor(s).
   c. Produce an assessment report to be delivered to the organization representative.

d. The assessment procedure ensures the confidentiality of the participants by signing a confidentiality agreement. The confidentiality of any data provided is guaranteed to all participants of the assessment.
   e. Feedback sessions are held after the assessment feedback is provided through a satisfaction questionnaire to be filled out by the sponsor. A post-mortem meeting is held between the assessors to discuss the performance of the assessment. Although this practice is evaluated to be fully adequate in the MARES method, at the end of the assessment, feedback session with the assessment participants like sponsors and organization representatives should be held to discuss not only the assessment results, but also to discuss, face to face, the assessment method and all issues addressed in the satisfaction questionnaire to get feedback directly, one should not rely only on the satisfaction questionnaire only.

4. **SPA documentation:**
   a. The assessment purpose, objectives and needed resources are all documented.
   b. Identification of the assessed organization unit.
   c. The confidentiality of the assessment is documented.
   d. The method documents all necessary templates and documents.
   e. Document the assessment process as a whole.
   f. Document of the assessment data and ratings.

5. **SPA users:**
   a. The responsibilities of the assessment participants are defined.
   b. The responsibilities of the assessment team members are defined.
   c. Senior management is involved in the assessment process.
   d. The participant commitment is ensured through the assessment briefing in the preparation phase.
   e. The method ensures the credibility of both sponsors and staff to believe that the assessment would give results.

## 4.2 Weaknesses

The evaluation criteria that are not fully met in the assessment method are considered as weaknesses. Therefore, based on the evaluation results, the following points are found to be weaknesses and need to be handled by the designers of the MARES method.

1. **SPA method:**
   a. Acquiring data through document review:
      MARES method optionally reviews documents to get data or to understand issues that are unclear in the interview. Although it has been indicated in [14] that for simple assessment, interviews or documents review is sufficient, we still believe that for accurate understanding both of them should be used.
   b. Studying the accuracy of findings.
      The collected data should be consolidated into accurate findings according to defined criteria. For the MARES method, no study has been found to define and test such accuracy of findings.

c. Conducting the assessment in 2-8 hours:
      The assessment tool indicated that the MARES method takes 2 days of assessment, analysis and presenting results: 2 days is justifiable for all these activities. The mentioned duration of 2-8 hours actually specifies the assessment time only, not the time for analysis and presenting results. Hence MARES method should bind itself to a specific assessment duration which should not exceed 8 hours.
   d. Ensure the reliability of the assessment results.
      Although the interview mentioned that reliability is ensured through numerous case studies, no study has been found to measure to what degree the assessment results produced by these case studies are repeatable. Ensuring repeatability will give confidence to the organization to rely on the assessment results and make further decisions for improvements.
   e. Ensure completeness.
      No study was found to discuss the completeness of the MARES method showing that the assessment method has taken into account the essential elements to assess each process in the assessment scope and give all needed results.

2. **SPA supportive tools:**
   So far, there is no tool support for the MARES method. A supportive tool that provides the following features is needed to achieve a successful assessment method:
   a. A tool that is usable and cover the different phases of the assessment.
   b. Create and use a database of historical assessment data.
   c. Build a database after the assessment process which contains the process profiles and other necessary data would be useful for new assessment trials and also for comparing assessment results with previous assessment trials results.
   d. Generating a semi-automatic assessment report.
      A supportive tool which produces parts of the final assessment report will produce a more efficient assessment process.
   e. Adaptable and flexible assessment tool.
      The supportive tool should allow adding or removing new processes to be assessedto fit the needs and goals of the assessed organization.

3. **SPA procedure:**
   The MARES method fulfills all the practices for the procedure class including:
   a. Prepare the assessment process by training participants, developing assessment plan and other related activities.
   b. Ensure formally the confidentiality of participants (even before conducting the assessment).
   c. Work to build confidence and trust relationships with participants
   d. Produce an assessment report to be delivered to the sponsors and organization.
   e. Hold a feedback session after each assessment to present results and get comments.

4. **SPA documentation:**
    a. Providing guidance for the assessment team.
       MARES should define the assessment team members' backgrounds, experience and their levels of knowledge in the assessment method and underlying model. This part of documentation also defines the team leader's management and technical skills related to the assessment process.
    b. Provide the guidance for follow-up meetings.
       The MARES method should provide guidance for assessors who will conduct the follow-up meetings after conducting the improvement phase.

5. **SPA user:**
    a. Ensuring that the sponsor and staff believe the assessment will give results.
       Another important aspect that would foster effective involvement in the assessment process is the belief that the assessment will give results and will lead to the improvement of the organization's behaviour and performance. The MARES method should work to build this belief before and during the assessment.
    b. Ensure that the benefits of the assessment method are felt by the participants.
       The assessment team of MARES method works on building such assurance during the preparation phase of the assessment. Being confident of the benefits would promote better involvement and cooperation on the part of the organization's participants in the assessment process.

By collecting statistics from the evaluation tool, as shown in Table 1 and summarized in Figure 1, the following can be noted:

1. The strengths related to the 'method' class gained the most contribution of the total strengths of the MARES method with a total contribution of about 26%, having about 77% of the practices in this class achieved.
2. The strengths related to the 'supportive tool' class gained the minimum contribution of the total strengths of the MARES method with a total contribution of 0%.
3. The strengths related to the 'documentation' class gained about 17% of the total strength contribution, achieving about 81% of the practices for this class.
4. The strengths related to the 'procedure' class gained about 13% of the total strength contribution, achieving 100% of the practices for this class.
5. The strengths related to the 'user' class gained about 13% of the total strength contribution, achieving 83% of the practices for this class.

For each category, the assessment method should satisfy at least 60% of the total practices for that class to be said that it has an acceptable level of success in that category. As shown by the data in Table 1, the 'supportive tool' class needs more work to achieve the minimum acceptable level of satisfaction.

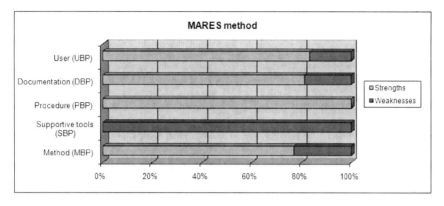

**Fig. 1.** MARES Assessment Method Evaluation Results

## 5 Conclusions and Future Works

The recent dissemination of number of process models and of their related assessment methods raises concerns related to their verification & validation (V&V), often not done, because it was supposed they were yet properly validated. Since the results from a process assessment provide the foundation and main input for writing an improvement plan, it is very critical for an organization to have confidence in its effectiveness for properly achieving the planned business results. In particular, a different scope must be designed when dealing with a SME or a VSE, because of their reduced organizational size and, consequently, some differences about the way to run and manage the business.

The evaluation criteria presented in [5] can provide useful information about strengths and weaknesses of software process assessment methods. In particular, five areas are covered, see Appendix A: Methods, Supportive Tools, Procedure, Documentation and Users. The information is sufficiently detailed to correctly guide an SPA designer on what to do and where – eventually – to correct the SPA method itself.

In order to start proving the relevance of our proposal, a first case study applying such evaluation criteria was done taking into account the MARES method, one of the well-known assessment methods for SME/VSE. The analysis revealed a number of candidate improvements to MARES.

This case study has illustrated how the proposed evaluation is useful in shedding light on the weaknesses of the evaluated assessment methods and providing guidance for the designers and users (assessors) of these methods. In the near future, the case study will be repeated evaluating further SPA methods: such case studies will help to refine more and more the above presented evaluation criteria. A step in this direction has been accomplished through an online survey, representing the evaluation tool used in this paper. The online survey has been published on line at http://www.kwiksurveys.com/online-survey.php?survey_ID=HLOJJO_9d452719. Until now, more that 30 responses have been collected.

## References

[1] López, M.: Application of an evaluation framework for analyzing the architecture tradeoff analysis method. Journal of Systems and Software 68, 233–241 (2003)
[2] Scriven, M.: Evaluation Thesaurus. Sage Publications, Newbury Park (1991)
[3] Shadish, W., Cook, T., Leviton, L.: Foundations of Program Evaluation. Theories of Practice. Sage Publications, Newbury Park (1993)
[4] Worthen, B., Sanders, J., Fitzpatrick, J.: Program Evaluation. Alternative Approaches and Practical Guidelines. Addison Wesley Longman, New York (1997)
[5] Zarour, M., Abran, A., Desharnais, J.M., Buglione, L.: Design and Implementation of Lightweight Software Process Assessment Methods: Survey of Best Practices. In: Proceedings of the 10th Software Process Improvement & Capability dEtermination conference (SPICE 2010), Pisa, Italy, pp. 39–50 (2010)
[6] Anacleto, A., Wangenheim, C., Salviano, C., Savi, R.: A method for Process Assessment in Small Software Companies. In: 4th International SPICE Conference on Process Assessment and Improvement, Portugal (2004)
[7] Wangenheim, C., Anacleto, A., Salviano, C.F.: MARES - A Methodology for Software Process Assessment in Small Software Companies, Technical Report: LQPS001_04E, LPQS - Universidade do Vale do Itajai, Brazil, Technical Report (2004)
[8] ISO/IEC, ISO/IEC 15504 Information Technology - Process Assessment - Parts 1-5 (2003-2006)
[9] Anacleto, A., Wangenheim, C.G., Salviano, C.F., Savi, R.: Experiences gained from applying ISO/IEC 15504 to small software companies in Brazil. In: 4th International SPICE Conference on Process Assessment and Improvement, Lisbon, Portugal (2004)
[10] Wangenheim, C., Varkoi, T., Salviano, C.F.: Standard based software process assessments in small companies. Software Process Improvement and Practice 11, 329–335 (2006)
[11] Wangenheim, C., Weber, S., Hauckc, J., Trentind, G.: Experiences on establishing software processes in small companies. Information and Software Technology 48, 890–900 (2006)
[12] Wangenheim, C.G., Anacleto, A., Salviano, C.F.: Helping small companies assess software processes. IEEE Software 23, 91–98 (2006)
[13] Wangenheim, C.G.v., Varkoi, T., Salviano, C.F.: Performing ISO 15504 Conformant Software Process Assessment in Small Software Companies. In: EUROSPI 2005, Hungary (2005)
[14] CMMI-Team, CMU/SEI-2006-TR-011: Appraisal Requirements for CMMI, Version 1.2 (ARC, V1.2), SCAMPI Upgrade Team, Carnegie Mellon Software Engineering Institute, Pittsburgh, PA CMU/SEI-2006-TR-011 (August 2006)

# Appendix A: The best practices grouped in five main classes

| | **Method related practices** |
|---|---|
| 1 | Does the method provide flexible and customizable method focusing on principal high-priority processes |
| 2 | Is the assessment method make use of simple, well-structured questionnaire with no more than 150 questions |
| 3 | Does the assessment method ensure, based on studies, the reliability of the assessment result |
| 4 | Does the method identify strengths, weaknesses, improvement opportunities and threats |
| 5 | Does the method suggest a feasible improvement action plan which addresses the special needs of the company |
| 6 | Does the collect data from interviews |
| 7 | Does the method check the accuracy of the assessment's findings (data collected) |
| 8 | Does the method comply with formal assessment method |
| 9 | Does the assessment take a reasonable time period |
| 10 | Does the method collect data from documents |
| 11 | Is the assessment method usable for on-site assessment and self-assessment |
| 12 | Does the method identify the process reference model used to select processes |
| 13 | Does the method ensure completeness |
| | **Supportive-tool related practices** |
| 1 | Does the assessment tool cover different assessment phases including collect, analyze and visualize data |
| 2 | Does the tool build and use a database of historical SPA data |
| 3 | Is the tool flexible/adaptable (i.e. adding new axes to the tool |
| 4 | Does the tool work to maintain assessment confidentiality (through security features for example) |
| 5 | Does the tool generate automatically assessment reports |
| 6 | Does the method work to ensure repeatability of the results |
| | **Procedure related practices** |
| 1 | Does the method prepare the assessment process by training participants, developing assessment plan and any related activities |
| 2 | Does the method hold a feedback session after each assessment to present results and get comments |
| 3 | Does the method ensure formally the confidentiality of participants (even before conducting the assessment) |
| 4 | Does the method work to build confidence and trust relationships with participants |
| 5 | Does the method produce an assessment report to be delivered to the sponsors and organization |
| | **Documentation related practices** |
| 1 | Does the method provide guidance that documents the assessment method and its implementation in practice |
| 2 | Does the method provide guidance in how to identify the assessment team needed skills |
| 3 | Does the method provide guidance that documents data collection and rating results |
| 4 | Does the method provide templates for the produced documents |
| 5 | Are the assessment objectives, purpose and needed resources documented? |
| 6 | Does the method provide guidance that ensures and highlights confidentiality terms |
| 7 | Does the method provide guidance that identifies the organizational unit |
| 8 | Does the method provide guidance for the follow-up assessors who will make reassessment later on |
| | **Users related practices** |
| 1 | Does the method ensure sponsors' commitment |
| 2 | Does the method ensure the involvement of senior management and other staff members in assessment |
| 3 | Are the assessment participants (interviewee) responsibilities defined? |
| 4 | Does the method define assessment team credentials and responsibilities |
| 5 | Does the method ensure that the benefits of the assessment are felt by participants |
| 6 | Does the method improve the credibility of both sponsor and staff who should believe that the assessment will yield a result |

# Functional Safety Extensions to Automotive SPICE According to ISO 26262

Per Johannessen[1], Öjvind Halonen[2], and Ola Örsmark[1]

[1] Volvo Car Corporation
pjohann1@volvocars.com, oorsmar1@volvocars.com
[2] EIS by Semcon
ojvind.halonen@eis.semcon.com

**Abstract.** The automotive industry is currently focused on feature development to deliver green, safe and connected vehicles. Implementations of these features increase both complexity and function integration in software as well as in electronic hardware. In order to maintain safety in vehicles due to this more complex and integrated environment, the upcoming ISO 26262 functional safety standard will give support. The automotive manufacturers who develop safety related functionality could benefit from using this new ISO standard to address functional safety. One requirement of ISO 26262 is to assess the capability of the development process used to comply with the standard. This paper describes an approach to extend ISO/IEC 15504 and Automotive SPICE to fulfill this ISO 26262 requirement for both software and hardware development. The functional safety extensions can be used together with Automotive SPICE for process assessments of functional safety in the automotive industry.

**Keywords:** Automotive SPICE, DFEA2020, Functional Safety, ISO/IEC 15504, ISO 26262, Safety Assessment.

## 1 Introduction

Today, the automotive industry is undergoing significant changes due to many different external factors, such as increased focus on environmental care, awareness of safety, and integration of consumer electronics in vehicles. At Volvo Cars, this is visible in feature development in the three key areas; green, safe and connected. This leads to an exponential growth of electronics and software in our vehicles. Together with an increasing focus on functional safety and dependability, there is a need to further develop both electrical architectures and development methods. To address these challenges in the automotive industry, the DFEA2020 national research project is conducted at Volvo Cars in collaboration with several partners. DFEA2020 is funded by VINNOVA, a Swedish government agency for innovation, and the DFEA2020 project partners.

Due to a new functional safety standard, ISO 26262 [1], one part of DFEA2020 is dedicated to functional safety and this standard. ISO 26262 is in its final stage before publication in 2011. Even if the standard has been used in its draft versions, the impact on the automotive industry for passenger cars will be significant. ISO 26262 is applicable for safety related functions, systems, and components that are implemented in electronics or software. ISO 26262 will address both safety related implementations and the development process used. For the development process, there is a need and also an ISO 26262 requirement to determine whether it is compliant with the process prescribed by ISO 26262.

A specific goal for DFEA2020 is to develop a framework for process assessment meeting the ISO 26262 standard. An interim result is described in this paper.

The paper starts with a brief overview of standards related to the proposed functional safety extensions and a summary of current state of functional safety assessments in the automotive industry. This is followed by a presentation of the selection of assessment framework and the proposed extensions. Further, two examples of the extensions as proposed to Automotive SPICE and ISO 15504 are included to give the reader a better understanding. Next, a guide to implementation is presented. Finally, the paper provides some conclusions and discusses ideas for further work.

## 2 Related Work

There are several standards and efforts to address functional safety in product development. Some of these are related to the proposed functional safety extensions to Automotive SPICE. The related standards are briefly introduced here.

### 2.1 ISO 26262

ISO 26262 [1] is the upcoming automotive standard for functional safety applicable for safety related Items that are implemented in electronics or software. An Item in [1] is defined as "system or array of systems to implement a function at the vehicle level, to which ISO 26262 is applied".

In the version to be released, the standard is limited to passenger cars up to 3.5 metric tonnes. Possibly, the standard could become applicable for heavy vehicles in its first revision.

The standard is currently available as Final Draft International Standard (FDIS), which is the last step before being a public international standard. ISO 26262 has been in development within ISO since 2005. From this time, it has increasingly been introduced for product development in the automotive industry.

The standard includes ten different parts and its lifecycle spans from concept development, through product development, to service, operation and decommissioning. With that scope, the impact from this standard on the automotive industry will be significant.

One key concept of ISO 26262 is the Automotive Safety Integrity Level (ASIL) which is determined during the concept phase of product development. ASIL is both a measure of risk for hazards and a measure of necessary risk reduction that should be addressed during product development. There are four levels, attributed ASIL A, ASIL B, ASIL C, and ASIL D where ASIL D implies the highest level of rigor during development. In case the Item has no hazard with an associated ASIL, a QM attribute is used. QM denotes Quality Management for which ISO 26262 has no requirements.

Many driveline and chassis systems need to manage ASIL C or ASIL D safety requirements while body systems often only have to manage ASIL A, ASIL B, or QM. In general, ASIL A and ASIL B systems do not typically require redundancy in electronic hardware which ASIL C and ASIL D typically do. Further, ISO 26262 has different rigor in requirements on the development process depending on the ASIL.

A Functional Safety Assessment shall ensure that Item under development has an appropriate level of functional safety according to ISO 26262, i.e. a functional safety product assessment. For an Item under development, this is done by checking that the Item has both the required documentation and the required safety measures implemented according to ISO 26262. Examples of required documentation are Hazard Analysis and Risk Assessment reports and Safety Cases. Typical safety measures are safety monitors and redundant sensors. The Functional Safety Assessment shall also consider the results of a Functional Safety Audit.

A Functional Safety Audit is required by ISO 262626 to ensure that the development process used for the Item is compliant with ISO 26262, i.e. a functional safety process assessment.

Both the Functional Safety Audit and the Functional Safety Assessment are depending on the ASIL of the Item under development. The higher the ASIL, the more processes are needed and the assessor should be more independent.

Even if ISO 26262 has requirements for Functional Safety Audits, there is no guidance in the standard on how to carry it out. However, the standard provides notes that the Safety Audit is related to SPICE assessments.

### 2.2 Automotive SPICE

The automotive industry in general has during the last decade focused on implementation of Automotive SPICE [2, 3], which is an adopted subset of ISO/IEC 15504 [4]. The vehicle manufacturers have strived to achieve process fulfillment at maturity level 3 for a subset of the processes required for their suppliers. During the last two years, the focus on Automotive SPICE has decreased and instead the focus has moved from quality to safety. Even if Automotive SPICE assessments are valuable for development of safety related software, there are several gaps to ISO 26262 that need to be addressed, e.g. system and hardware development.

### 2.3 +SAFE

+SAFE [5] is an extension to Capability Maturity Model Integration (CMMI) for Development (CMMI-DEV) that covers safety management and safety engineering. It was developed by the Defence Materiel Organisation within the Australian Department of Defence and the latest version, version 1.2, was released in 2007. The +SAFE extension supplements CMMI-DEV with two additional process areas that provide a basis for appraising or improving an organization's processes for providing safety-critical products.

+SAFE was developed for standalone use. It is not intended to be embedded in a CMMI model, but can be modified to support different safety standards.

This extension is a good starting point for ISO 26262 process capability determination. However, +SAFE is not sufficient by itself to be used in the automotive industry due to gaps to ISO 26262, e.g. electronic hardware processes are missing in +SAFE.

## 2.4 ISO/IEC 15504-10 - Safety Extension

This part 10 of ISO/IEC 15504 [6] defines processes and guidance to support the development of safety related systems. It is currently under development and is expected to be released during 2011. The process assessment model for this part 10 complements the process assessment model for system and software as defined in ISO/IEC 15504 Parts 5 and 6.

ISO/IEC 15504-10 defines three processes to support safety. The processes are:

- Safety Management process
- Safety Engineering process
- Safety Qualification process

ISO/IEC 15504-10 claims that the defined processes are consistent with the five different safety standards:

- IEC 61508
- +SAFE, A Safety Extension to CMMI-DEV, V.1.2.
- IEC 60880
- UK MoD Def Stan 00-56
- ISO 26262

These five standards use different safety lifecycles with different processes and it is challenging to write a standard such as ISO/IEC 15504-10 to cover all of these processes. In the case of ISO 26262, there are gaps between the three additional processes defined in ISO/IEC 15504-10 and the processes needed to be assessed according to ISO 26262, e.g. electronic hardware processes are missing in ISO/IEC 15504-10. In this paper, we suggest additional processes to close this gap.

## 2.5 Functional Safety Assessments in the Automotive Industry

Currently in the automotive industry, there is no commonly used functional safety assessment framework. To a large degree, functional safety assessments are done by expert judgment. Some companies, offering functional safety assessment, use company internal instructions and checklist when doing these assessments for customers. However, from our perspective, there is currently a large degree of ad-hoc functional safety assessments performed in the automotive industry. This is particular true when it comes to functional safety processes assessments, for processes not covered by Automotive SPICE. Further, most functional safety assessments seem to focus more on technology rather than development processes used. As ISO 26262 soon will be released as an international standard, there is a need in the industry to standardize functional safety process assessments.

## 2.6 Swedish Standardization for Functional Safety Process Extensions

In the Swedish working group for ISO 26262, a task has been initiated to develop a Swedish standard based on the functional safety process extensions described in this paper. Sweden has four major automotive manufacturers with a strong tradition on safety, all participating in this standardization effort.

The goal is to have an international standard instead of a Swedish national standard. The main reasons for developing a Swedish standard is that it will be a good basis to propose as an international standard and it could be developed in less than six months. With the current plan for this standard, it would be available approximately at the same time as ISO 26262 is released as an international standard. Hence, this Swedish standard could be used by any early adopter and be a proposal for a new international standard.

# 3 Functional Safety Process Assessment Strategy

In order to ensure that the development of safety related systems will result in safe systems, an assessment strategy is needed. Two parts have been identified as necessary for functional safety process assessment:

- Performing a functional safety process assessment on a reference project before development, i.e. a process capability determination.
- Performing a tailored functional safety process assessment during development as a part of the functional safety product assessment.

The strategy, as can be seen in Fig 1, implies a wider context than ISO 26262 suggests where functional safety process assessment is only required to be performed on projects during development. The possibility to assess a reference project adds significant value, e.g. at procurement, and supports long term relations with suppliers. During development the effort will be reduced accordingly.

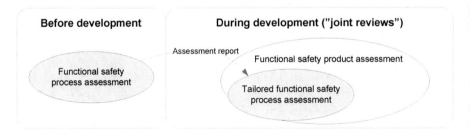

**Fig. 1.** Overall strategy for functional safety process assessment

The ability to tailor the assessments for different ASIL becomes essential since the ISO 26262 requirements depend on the ASIL. The selected approach is to base the assessment framework on ASIL D requirements and allow for further tailoring by the assessors. Adapting this assessment framework to each ASIL would be impracticable.

## 4 Assessment Framework

Implementation of the functional safety process assessment strategy will benefit from a supporting assessment framework. An assessment framework requires a process reference model to scope the ISO 26262 objectives, requirements, and work products, as well as a process assessment model.

### 4.1 Selection of Assessment Framework

Three different approaches for developing the assessment framework were considered:

- Use the Automotive SPICE framework
- Develop custom framework based on ISO 26262
- Use framework from other domains

The first approach, reusing the Automotive SPICE framework, was found to be quite attractive due to the organizational knowledge, established methods for SPICE assessment, and global knowledge of SPICE assessment in the industry. However, an extended framework for functional safety assessment would need to be developed.

The second approach, to develop a custom framework based upon ISO 26262, was attractive due to the lightweight approach, to just use what was required, and easy to tailor for its needs. However, this approach would be informal and hard to establish especially for suppliers since it would not relate to established assessment practices. To compare the results from different assessments would also be hard with a custom framework. This approach is similar to the ad-hoc approach taken by assessment companies today as discussed earlier in this paper.

To use a framework from other domains would give the benefit of an established framework and certification scheme. One such framework is CASS (Conformity Assessment of Safety-related Systems) [7] which is based on the standard IEC 61508 [8]. However, major adoption to automotive industry and ISO 26262 would be required.

The decision was to reuse the Automotive SPICE assessment framework, extended by ISO 26262 compatible processes for functional safety process assessment. This gave most organizational and domain benefits.

### 4.2 Assessment Model

Four categories of processes have been identified with respect to Automotive SPICE, as shown in Fig. 2, in order to incorporate the ISO 26262 requirements:

- Additions for functional safety to Automotive SPICE
- Additions for functional safety to ISO/IEC 15504
- New processes, also called extensions according to SPICE, to Automotive SPICE and ISO/IEC 15504 for unique functional safety processes
- Reused processes, which are processes in Automotive SPICE that need to be assessed to show full compliance with ISO 26262

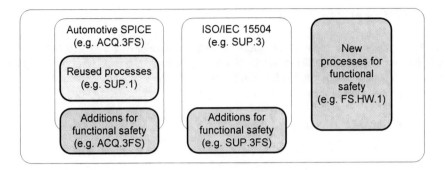

**Fig. 2.** Process extensions for functional safety process assessment

The functional safety process assessment framework extends the Automotive SPICE model by adding new processes or making additions to existing processes. Existing process outcomes were not altered and none were removed.

### 4.3 Identification of Extensions

A process ID is used for identification of the functional safety extensions, but in order to separate those from the Automotive SPICE processes, the processes and practices have been uniquely defined by adding FS, representing Functional Safety.

Additions are identified through a postfix, ".FS" to the process ID. As an example, the functional safety addition to the Automotive SPICE process ACQ.3, Contract Agreement, is identified as ACQ.3FS. As a second example, Software Validation which is only defined according to ISO/IEC 15504 has to be extended with safety validation, which is identified as the addition SUP.3FS.

The new processes, i.e. extensions, are identified through a prefix, "FS" to the process ID. As an example, Automotive SPICE does not cover hardware development and a new process needs to be developed. The new process is identified as FS.HW.1.

Note that reused processes have been included but without the FS prefix/postfix. The process definitions for these processes are not modified, but are essential to meet the assessment scope of ISO 26262.

In order to get a complete process assessment model, each safety related process has been extended with so-called safety practices (SP) to cover the wider scope of ISO 26262. The safety practices correspond to base practices in Automotive SPICE.

Safety practices are identified through the process extensions ID and by adding a postfix ".SP#", e.g. SUP.3FS.SP1

### 4.4 Functional Safety Extensions to Automotive SPICE and ISO/IEC 15504

The required functional safety extensions, and their associated work products, have been established and traced to ISO 26262. The four tables below show the resulting functional safety extensions, including the reused processes, needed to cover ISO 26262 requirements.

**Table 1.** Overview of functional safety additions to existing Automotive SPICE processes

| Automotive SPICE additions | Process name | ISO/DIS 26262 references |
|---|---|---|
| MAN.3FS | Project Management | 2-5, 8-5, 4-5 |
| SPL.2FS | Product release | 4-11 |
| SUP.10FS | Change request management | 3-6, 8-8 |
| ENG.2FS | System requirements analysis | 3-5, 4-6, 8-6 |
| ENG.3FS | System architectural design | 4-7 |
| ENG.4FS | Software requirements analysis | 6-6 |
| ENG.5FS | Software design | 6-7 |
| ENG.6FS | Software construction | 6-8, 6-9 |
| ENG.7FS | Software integration test | 6-10 |
| ENG.8FS | Software testing | 6-10, 8-9 |
| ENG.9FS | System integration test | 4-8 |
| ENG.10FS | System testing | 4-8, 6-11, 8-9 |
| ACQ.3FS | Contract Agreement | 8-5 |

**Table 2.** Overview of functional safety additions to existing ISO/IEC 15504 processes

| ISO/IEC 15504 additions | Process name | ISO/DIS 26262 references |
|---|---|---|
| SUP.3FS | Validation | 4-5, 4-6, 4-9 |

**Table 3.** Overview of new functional safety processes, i.e. extensions, to Automotive SPICE and ISO/IEC 15504

| Extension | Process name | ISO/DIS 26262 references |
|---|---|---|
| FS.MAN.1 | Safety Culture Management | 2-5 |
| FS.MAN.2 | Safety Life Cycle Management | 2-5, 2-6, 2-7, 3-6 |
| FS.AN.1 | Hazard Analysis | 3-7 |
| FS.AN.2 | Safety Analysis on System Level | 4-7, 9-7, 9-8 |
| FS.AN.3 | Hardware Safety Analysis | 5-7, 5-8, 5-9, 9-7, 9-8 |
| FS.AN.4 | Software Safety Analysis | 6-7, 9-7, 9-8 |
| FS.SUP.1 | Safety Case Development | 2-6, 8-10 |
| FS.SUP.2 | SW Component Qualification | 8-12 |
| FS.SUP.3 | HW Component Qualification | 8-13 |
| FS.SUP.4 | Calibration and Configuration Data Management | 6-Annex C |
| FS.HW.1 | Hardware Safety Engineering | 5 |
| FS.TOOL.1 | Qualification of Tools | 8-11 |
| FS.PROD.1 | Production, Operation, Service and Decommissioning | 7 |

**Table 4.** Overview of reused processes from existing Automotive SPICE processes

| Automotive SPICE | Process name | ISO/DIS 26262 references |
|---|---|---|
| MAN.5 | Risk management | - |
| SUP.1 | Quality assurance | 2-5 |
| SUP.2 | Verification | 4-5, 4-8, 4-9, 5-10, 6-9, 6-10, 6-11, 8-9 |
| SUP.4 | Joint review | 8-5 |
| SUP.8 | Configuration management | 8-7 |
| SUP.9 | Problem resolution management | 2-4 |
| REU.2 | Reuse program management | - |
| ACQ.4 | Supplier monitoring | 8-5 |

## 4.5 Examples of Extensions and Safety Practices

As described there are different types of extensions needed. For these, it was decided to use the same structure as in Automotive SPICE.

In Table 5, there is an example of an extension to an Automotive SPICE process. This extension is an addition that is needed for change management since the SUP.10 process does not include safety analysis, safety lifecycle, and safety manager approval. Therefore, these three aspects have to be added to SUP.10 as shown in Table 5.

**Table 5.** The Change Request Management addition to SUP.10 in Automotive SPICE

| Process ID | SUP.10FS | Applicable ASIL | A-D |
|---|---|---|---|
| **Process Name** | Change request management | | |
| **Process Purpose** | The purpose of the FS addition to the change request management process is to ensure that safety related work products are analyzed and managed during the entire safety lifecycle. | | |
| **Process Outcomes** | As a result of successful implementation of this process: 1) the change request is analyzed for impact on functional safety of the product; 2) the change request is analyzed for impact of the safety lifecycle and what safety activities that needs to be carried out again; 3) if there are changes related to safety, the decision to accept, reject or delay the change is agreed with the safety manager. | | |
| **Safety Practices** | SUP.10FS.SP1: Perform an impact analysis with respect to safety of the product. This analysis should also include new or changed hazards. SUP.10FS.SP2: Perform an impact analysis with respect to the functional safety activities that need to be conducted. If the ASIL level is increased because of the change, a gap analysis shall be performed to find out what needs to be done to achieve the higher ASIL. SUP.10FS.SP3: The safety manager is included in the decision process. | | |

There are five different work products in ISO 26262 impacted by SUP10.FS, these are shown in Table 6 together with their reference to ISO 26262.

**Table 6.** The work products from the Change Request Management extension SUP.10FS with the corresponding ISO 26262 references

| Output Safety Work Products | ISO/DIS 26262 reference |
|---|---|
| Change management plan | 8-8.5.1 |
| Change request | 8-8.5.2 |
| Impact analysis | 8-8.5.3 |
| Change request plan | 8-8.5.3 |
| Change report | 8-8.5.4 |

Another type of extensions to Automotive SPICE and ISO/IEC 15504 is for new processes. An example of this extension is shown in Table 7, which is for hazard analysis. The hazard analysis process is one key process needed for functional safety as the whole safety lifecycle is impacted by the outcome of this process.

**Table 7.** The Hazard Analysis extension as proposed

| Process ID | FS.AN.1 | Applicable ASIL | QM, A-D |
|---|---|---|---|
| **Process Name** | Hazard Analysis | | |
| **Process Purpose** | The purpose of the hazard analysis is to identify and classify hazards related to the item. | | |
| **Process Outcomes** | As a result of successful implementation of this process: <br> 1) the target for the analysis is clearly defined; <br> 2) the failure modes of actuators and functions (use cases) are identified; <br> 3) the relevant situations are identified; <br> 4) the hazards are clearly expressed; <br> 5) the hazards are classified according to an international standard; <br> 6) the top level safety requirements (safety goals) are clearly defined; <br> 7) the analysis is revised during the development to seek for new or changed hazards. | | |
| **Safety Practices** | FS.AN.1.SP1: Define the item. This includes external interfaces, functional requirements, non-functional requirements, assumptions and foreseeable misuse. <br><br> FS.AN.1.SP2: Define the failure modes of 1) actuators and 2) functions (e.g. based on use cases). Define the system effect (technical) and the effect on the vehicle level as a result of the failure modes. *NOTE: Omission (no effect) and commission (full effect when not wanted) shall always be considered. Other failure modes (late, early, more, less and stuck effect) may be considered depending on the function.* <br><br> FS.AN.1.SP3: Define the relevant situations for the failure modes. This includes all common operating situations (e.g. driving on straight road, city driving, and situation when function is used) and may also include production, service and other special situations. The situation coverage should be determined. <br><br> FS.AN.1.SP4: Define the hazards based on the failure modes and situations. <br><br> FS.AN.1.SP5: Perform classification according to an international standard relevant for the automotive industry. Justify the classifications with descriptions of the assumptions made. Involve a group of people in the hazard analysis effort. <br><br> FS.AN.1.SP6: Define the top level safety requirements (safety goals) together with their safety integrity levels and safe states. <br><br> FS.AN.1.SP7: Revise the analysis during development. | | |

There are three different work products in ISO 26262 impacted by FS.AN.1, these are shown in Table 8 together with their reference to ISO 26262.

**Table 8.** The work products from the Hazard Analysis extension FS.AN.1 with the corresponding ISO 26262 references

| Output Safety Work Products | ISO/DIS 26262 reference |
|---|---|
| Hazard analysis and risk assessment | 3-7.5.1 |
| Safety goals | 3-7.5.2 |
| Verification review of hazard analysis and risk assessment and safety goals | 3-7.5.3 |

## 5 Implementation

When doing functional safety assessments on Items according to ISO 26262 and the proposed work in this paper, there are four steps recommended:

- Request an Automotive SPICE Assessment from the supplier
- Carry out a functional safety process assessment on a reference project, decided jointly with the supplier, based on the assessment framework presented in this paper
- Perform a functional safety product assessment on the Item during the product development, from project start to start of series production
- Follow up the action plan from the functional safety process assessment and the functional safety product assessment during the project.

For functional safety process assessment, maturity level 0-3 should be assessed where level 3 is required. Level 3 should be sufficient since ISO 26262 does not require organizational process implementation and level 3 is commonly accepted as a minimum level for Automotive SPICE compliance. Level 3 has the advantage that the outcome should have minimum dependency on specific development projects. Further, it is also possible to tailor the process assessment depending on the ASIL. The ASIL to be used could for instance depend on the types of systems developed by the assessed organization.

For functional safety product assessment, maturity level 0-1 is assessed for work products, where level 1 is required. The reason for choosing level 1 is that this type of assessment is focused on the achievement of the process purpose, i.e. safety practices, and the characteristics of the work products. To ensure that the development process used for the Item is compliant with ISO 26262, it is sufficient to check that the development process has satisfactorily been assessed and, by simple checks, confirm that the previously assessed process is being used in the development. Inspection of the work products refers to the same processes as used for the tailored functional safety process assessment.

## 6 Conclusion and Further Work

The work to implement safety extensions to ISO/IEC 15504 and Automotive SPICE was quite straight forward once the methodology for the extensions was set. Further, since the ISO 26262 requirements on safety organizations are similar to Automotive

SPICE requirements, the smoothest approach was to reuse the software quality organization setup and the software process assessment methods for the ISO 26262 safety organization and functional safety assessments.

The challenge has been to verify the extensions through product and process assessments. This work is ongoing within the DFEA2020 project and also in vehicle programs within Volvo Cars.

Since quality and safety go hand in hand, the people working with safety assessment and those working with quality assessment may well be in the same organization. If similar work methods can be used, higher efficiency will be achieved at an organizational level. Tools already in use for SPICE assessment can be expanded to also support the extensions required for functional safety assessment.

For organizations which base their processes on ISO/IEC 15504 and not Automotive SPICE, the safety extensions presented in this paper should be easy to adopt.

Once the Swedish working group for ISO 26262 has developed a Swedish standard for the process extensions described in this paper, the next step, apart from its use in product development, will be to target international standardization.

## Acknowledgments

This work has been funded by the Swedish FFI program (Strategic Vehicle Research and Innovation) through the project DFEA 2020.

The authors would like to thank Joacim Bergman who participated in the work developing the Safety Extensions presented in this paper.

## References

1. ISO/DIS 26262, Road vehicles - Functional safety, International Organization for Standardization, Geneva, Switzerland (2009)
2. Automotive SPICE Process Reference Model, v4.5, Automotive SIG (2010)
3. Automotive SPICE Process Assessment Model, v2.5, Automotive SIG (2010)
4. ISO/IEC 15504:2006 Information Technology - Process Assessment – Part 5: An exemplar Process Assessment Model. International Organization for Standardization, Geneva, Switzerland (2006)
5. +SAFE, A Safety Extension to CMMI-DEV, V1.2, Defence Materiel Organisation. Australian Department of Defence, Software Engineering Institute, Carnegie Mellon University, Pittsburgh, PA, USA (2007)
6. ISO/IEC DTR 15504-10, Information technology – Software process assessment – Part 10: Safety Extensions. International Organization for Standardization, Geneva, Switzerland (2010)
7. The CASS Guide to Functional Safety Management Assessment, Issue 2.a. The CASS Scheme Ltd., United Kingdom (2000)
8. IEC 61508, Functional safety of electrical/electronic/programmable electronic safety-related systems. International Electrotechnical Commission, Geneva, Switzerland (1998, 2000)

# An ISO/IEC 15504 Security Extension

Antoni Lluís Mesquida, Antònia Mas, and Esperança Amengual

University of the Balearic Islands, Department of Mathematics and Computer Science,
Cra. de Valldemossa, km 7.5, 07122 Palma de Mallorca (Illes Balears), Spain
{antoni.mesquida,antonia.mas,eamengual}@uib.es

**Abstract.** Software companies which have been involved in a process improvement programme according to ISO/IEC 15504 have already performed some steps in order to implement ISO/IEC 27000 as an information security management framework. After analysing in depth the existing relations between ISO/IEC 15504-5 base practices and ISO/IEC 27002 security controls, in this paper the security controls covered by the ISO/IEC 15504-5 processes are described, the changes over these processes which would be necessary for the implementation of the controls are detailed and an ISO/IEC 15504 Security Extension that facilitates the implementation of both standards is presented.

**Keywords:** ISO/IEC 15504 (SPICE), ISO/IEC 27000, Information security, Software Process Improvement (SPI).

## 1 Introduction

For software development companies the implementation of information security controls is fundamental to assure their continuity, minimise possible injuries and maximize the return of investment and business opportunities.

Within our environment, several software companies that have been involved in a process improvement programme according to ISO/IEC 15504 [1][2] demand the implementation of ISO/IEC 27000 as a security standard. In fact, some of these software organizations have been certificated against the ISO/IEC 27001 standard [3]. In order to obtain this certification, an organization must adequately select and implement the appropriate security controls between all the controls provided by the ISO/IEC 27002 standard [4].

Heads of quality departments in these organizations observed that some of the actions previously performed to implement the ISO/IEC 15504 standard were very useful when time came to implement the ISO/IEC 27001 standard. Therefore they could take advantage of their experience in the implementation of ISO/IEC 15504 base practices in order to implement the selected ISO/IEC 27002 security controls.

MiProSoft, our research group, is experienced in implementing ISO/IEC 15504 in software companies [5][6][7][8]. Moreover we have also analysed the relationship between this standard with other ISO standards, such as ISO 9001 [9][10] and ISO/IEC 20000 [11]. In this article we focus in the relationship between the security aspects of ISO/IEC 27000 series to the base practices of ISO/IEC 15504-5 [12] with a double objective:

- Facilitating the implementation of ISO/IEC 27001 in organizations which have already reached a particular maturity level according to ISO/IEC 15504-7 [13].
- Defining a method for the implementation of ISO/IEC 15504 which already considers the ISO/IEC 27001 security aspects.

After a complete analysis of all the existing relations between ISO/IEC 15504-5 processes and ISO/IEC 27002 security controls, it can be stated that ISO/IEC 15504-5 considers an important number of the security aspects and controls which are necessary for the implementation of an Information Security Management System [14]. Consequently, software companies which have been involved in a process improvement programme according to this standard have already performed some steps in order to implement ISO/IEC 27000.

Delving into this result, from the existing relations between the ISO/IEC 27002 security controls and the ISO/IEC 15504-5 base practices, and after analysing in depth the purpose and requirements of each security control and the related base practices, in this paper the adaptations and modifications that should be done in ISO/IEC 15504-5 processes in order to include the security aspects of the ISO/IEC 27002 related controls are presented.

The title of the paper "An ISO/IEC 15504 Security Extension" must be understood as an adaptation or amplification of ISO/IEC 15504-5 processes. In this work, an expansion of the ISO/IEC 15504-5 process map, like the one in ISO/IEC PDTR 15504-10 Safety Extensions [15], is not proposed.

The paper is structured as follows: Section 2 presents a summary of the ISO/IEC 27002 security controls addressed by the ISO/IEC 15504-5 processes. Section 3 describes the different kind of actions that could be performed over an ISO/IEC 15504-5 process to adapt it in order to satisfy all the security requirements of the related controls. Section 4 presents the Security Extension. Finally, Section 5 concludes this paper and opens discussions regarding the results.

## 2 ISO/IEC 27002 Security Controls Covered by ISO/IEC 15504

The controls covered by ISO/IEC 15504-5 processes are next presented using the Organizational Maturity Model defined in ISO/IEC 15504-7 which describes a framework for determining organizational maturity. Table 1 shows the 17 ISO/IEC 27002 security controls covered by the ISO/IEC 15504-7 Maturity Level (ML) 1 processes whereas Table 2 shows the 29 security controls covered by the ML 2 processes.

Table 1. ISO/IEC 27002 controls covered by the ISO/IEC 15504-7 ML 1 processes

| ISO/IEC 15504-7 ML 1 processes | ISO/IEC 27002 controls covered by the ML1 processes |
|---|---|
| ENG.1 Requirements elicitation | 10.1.4 Separation of development, test and operational facilities |
| ENG.2 System requirements analysis | 10.8.3 Physical media in transit |
| ENG.3 System architectural design | 10.9.1 Electronic commerce |
| ENG.4 Software requirements analysis | 10.9.2 On-line transactions |
| | 10.9.3 Publicly available information |

**Table 1.** (*continued*)

| ISO/IEC 15504-7 ML 1 processes | ISO/IEC 27002 controls covered by the ML1 processes |
|---|---|
| ENG.5 Software design | 12.1.1 Security requirements analysis and specification |
| ENG.6 Software construction | 12.2.1 Input data validation |
| ENG.7 Software integration | 12.2.2 Control of internal processing |
| ENG.8 Software testing | 12.2.3 Message integrity |
| ENG.9 System integration | 12.2.4 Output data validation |
| ENG.10 System integration | 12.4.1 Control of operational software |
| ENG.11 Software installation | 12.4.2 Protection of system test data |
| ENG.12 Software and system maintenance | 12.5.2 Technical review of applications after operating system changes |
| SPL.2 Product release | 15.1.1 Identification of applicable legislation |
|  | 15.1.2 Intellectual property rights (IPR) |
|  | 15.1.4 Data protection and privacy of personal information |
|  | 15.1.6 Regulation of cryptographic controls |

**Table 2.** ISO/IEC 27002 controls covered by the ISO/IEC 15504-7 ML 2 processes

| ISO/IEC 15504-7 ML2 processes | ISO/IEC 27002 controls covered by the ML2 processes |
|---|---|
| SUP.1 Quality Assurance | 6.2.1 Identification of risks related to external parties |
| SUP.2 Verification | 6.2.3 Addressing security in third party agreements |
| SUP.3 Validation | 10.1.1 Documented operating procedures |
| SUP.4 Joint review | 10.1.2 Change management |
| SUP.7 Documentation | 10.2.1 Service delivery |
| SUP.8 Configuration Management | 10.2.2 Monitoring and review of third party services |
|  | 10.2.3 Managing changes to third party services |
| SUP.9 Problem Resolution Management | 10.3.2 System acceptance |
|  | 10.5.1 Information back-up |
| SUP.10 Change Request Management | 10.6.2 Security of network services |
|  | 10.7.3 Information handling procedures |
| MAN.3 Project Management | 10.7.4 Security of system documentation |
| MAN.5 Risk Management | 10.8.2 Exchange agreements |
| ACQ.3 Contract Agreement | 12.4.3 Access control to program source code |
| ACQ.4 Supplier Monitoring | 12.5.1 Change control procedures |
| ACQ.5 Customer Acceptance | 12.5.3 Restrictions on changes to software packages |
| SPL.3 Product Acceptance Support | 12.5.5 Outsourced software development |
|  | 12.6.1 Control of technical vulnerabilities |
|  | 13.1.1 Reporting information security events |
|  | 13.1.2 Reporting security weaknesses |
|  | 13.2.1 Responsibilities and procedures |
|  | 13.2.2 Learning from information security incidents |
|  | 13.2.3 Collection of evidence |
|  | 15.1.2 Intellectual property rights (IPR)* |
|  | 15.1.3 Protection of organizational records |
|  | 15.1.4 Data protection and privacy of personal information* |
|  | 15.1.6 Regulation of cryptographic controls* |
|  | 15.2.1 Compliance with security policies and standards |
|  | 15.2.2 Technical compliance checking |

In Table 2 the controls marked with an asterisk have also been considered by the ML 1 processes. However, these controls are also listed in Table 2 because some processes of ML 2 cover new security requirements not considered before.

Finally, the ML 3 processes cover 57 new security controls. Furthermore, and like before, they also cover other security requirements of 11 controls already considered by processes of maturity levels 1 and 2. The controls covered by the ML 3 processes are not listed in this paper due to space requirements.

Given the above, the processes of ISO/IEC 15504-7 maturity levels 1, 2 and 3 could be used to get to satisfy 100 of the 133 ISO/IEC 27002 security controls.

## 3 ISO/IEC 15504-5 Adaptation to Satisfy ISO/IEC 27002 Controls

After analysing in depth the relations presented in the former section, in this section the modifications and amplifications that should be done in ISO/IEC 15504-5 processes in order to make them compliant with the security requirements of the related ISO/IEC 27002 controls are detailed. These changes could affect to different process components: process purpose, base practices or/and work products.

It is possible to establish four different kinds of actions to be performed on an ISO/IEC 15504-5 process in order that it covers a specific security control:

- Use the process purpose or its base practices to manage the security requirements of the related control, without any kind of process modification.
- Modify or extend one or more base practices.
- Add a new base practice from the related control objective, closely linked to the already existent base practices.
- Modify or extend the process purpose.

Subsections 3.1 to 3.4 show some examples of these four possible kinds of actions.

### 3.1 Using ISO/IEC 15504-5 Base Practices to Satisfy Security Controls

In this first case, ISO/IEC 15504-5 base practices can be directly used to satisfy the security aspects of the related control. No process modification or amplification needs to be done to completely cover the security control.

One example of this case can be observed in the connection between the *10.1.2 Change Management* control ("Changes to information processing facilities and systems should be controlled") and the *SUP.10 Change request management* process whose purpose is "to ensure that changes to products in development are managed and controlled". The SUP.10 base practices BP1 to BP9 can be performed in order to manage changes to information processing facilities in the manner indicated by the related control.

Another example of this case is the relation between the *10.1.1 Documented operating procedures* control ("Operating procedures should be documented, maintained, and made available to all users who need them") and the *SUP.7 Documentation* process whose purpose is "to develop and maintain the recorded information produced by a process". The SUP.7 base practices BP1 to BP8 can be performed in order to develop, maintain and make available the operating procedures.

## 3.2 Extending ISO/IEC 15504-5 Base Practices

In this case, the related ISO/IEC 15504-5 base practices could be widened to cover all the security aspects of the control.

One example of this case can be observed in the relation between the *13.1.1 Reporting information security events* control ("Information security events should be reported through appropriate management channels as quickly as possible") and the *RIN.3.BP4 Capture knowledge* ("Identify and record each knowledge item according to the classification schema and asset criteria"). In order to cover all the security aspects of the control, the description of *RIN.3.BP4* could be widened with the underlined sentence: "Identify and record each knowledge item according to the classification schema and asset criteria, including information security events through appropriate management channels as quickly as possible".

Another example of this case is in the connection between the *8.2.2 Information security awareness, education, and training* control ("All employees of the organization and, where relevant, contractors and third party users should receive appropriate awareness training and regular updates in organizational policies and procedures, as relevant for their job function. Ongoing training should include security requirements, legal responsibilities and business controls, as well as training in the correct use of information processing facilities e.g. log-on procedure, use of software packages and information on the disciplinary process") and the *RIN.2.BP2 Identify needs for training* ("Identify and evaluate skills and competencies to be provided or improved through training"). As the previous example, the description of this base practice could be widened to state: "Identify and evaluate skills and competencies to be provided or improved through training, including security requirements, legal responsibilities and business controls, as well as training in the correct use of information processing facilities".

## 3.3 Adding a New Base Practice

In this case, the related ISO/IEC 15504-5 process does not have any specific base practice that covers the security control and, therefore, it is necessary to create a new one.

One example can be observed in the *10.7.4 Security of system documentation* control ("System documentation should be protected against unauthorized access"), which is related to the *SUP.7 Documentation* process purpose: "To develop and maintain the recorded information produced by a process". In order to cover the security aspects of the control, the description of the new base practice, called *SUP.7.BP9 Protect documents*, should be: "Protect system documentation against unauthorized access".

A second example is the case of the *12.4.2 Protection of system test data* control ("Test data should be selected carefully, and protected and controlled. If personal or otherwise sensitive information is used for testing purposes, all sensitive details and content should be removed or modified beyond recognition before use"). This control is related to the *ENG.8 Software testing* process whose purpose is "to confirm that the integrated software product meets its defined requirements". In this case, a new base practice has been created: *ENG.8.BP0 Protect test data* ("Remove or modify beyond recognition before use, protect and control all personal or sensitive information used for testing purposes").

## 3.4 Adapting the Purpose of an ISO/IEC 15504 Process

In this case, there is a correspondence between a control and a process without an explicit connection with a particular base practice of the process. The relation has been identified by comparing the control description with the process purpose. Consequently, the action to be performed should consist on modifying or expanding the process purpose in order to cover the security requirements of the control.

This is the case of the *10.7.4 Security of system documentation* control ("System documentation should be protected against unauthorized access") with the *SUP.7 Documentation* process: "To develop and maintain the recorded information produced by a process". To cover the security aspects the process purpose has been changed to: "To develop, maintain and protect against unauthorized access the recorded information produced by a process".

Another example of this case can be observed in *ACQ.3 Contract agreement* process. *ACQ.3* is related to nine different ISO/IEC 27002 security controls, which are shown in Table 3.

Table 3. ISO/IEC 27002 controls covered by the ACQ.3 Contract agreement process

| ISO/IEC 27002 controls covered by the ACQ.3 process |
|---|
| 6.2.1 Identification of risks related to external parties |
| 6.2.3 Addressing security in third party agreements |
| 10.2.1 Service delivery |
| 10.6.2 Security of network services |
| 10.8.2 Exchange agreements |
| 12.5.5 Outsourced software development |
| 15.1.2 Intellectual property rights (IPR) |
| 15.1.4 Data protection and privacy of personal information |
| 15.1.6 Regulation of cryptographic controls |

In order to satisfy these controls, contracts or agreements negotiated and approved according to ACQ.3 should be widened, including specific clauses:

- involving accessing, processing, communicating or managing the organization's information or information process facilities (to satisfy controls *6.2.1* and *6.2.3*)
- obliging third parties to implement, operate and maintain the security controls, service definitions and delivery levels agreed (to satisfy control *10.2.1*)
- including all the security features, service levels, and management requirements of all network services (to satisfy control *10.6.2*)
- treating the exchange of information and software between the organization and external parties (to satisfy control *10.8.2*)
- including relevant aspects related to licensing arrangements, code ownership, intellectual property rights, rights of access for audit of the quality and accuracy of work done and contractual requirements for quality and security functionality of code, when software development is outsourced (to satisfy control *12.5.5*)

- ensuring the compliance with legislative, regulatory, and contractual requirements on the use of material in respect of which there may be intellectual property rights and on the use of proprietary software products (to satisfy control *15.1.2*)
- ensuring data protection and privacy as required in legislation and regulation agreed (to satisfy control *15.1.4*)
- referring to the cryptographic controls that should be used in compliance with all relevant agreements, laws, and regulations agreed (to satisfy control *15.1.6*)

In this way, if an organization which has implemented the *ACQ.3* base practices adds to the standard contract the former clauses, it would have also implemented the nine ISO/IEC 27002 security controls related to this process.

## 4 ISO/IEC 15504 Security Extension

The ISO/IEC 15504 Security Extension has a double application. More concretely, it could be used to:

- Facilitate the implementation of the ISO/IEC 27001 standard in software organizations which are or have been involved in SPI programmes according to ISO/IEC 15504.
- Facilitate the simultaneous implementation of both ISO/IEC 27001 and ISO/IEC 15504 standards, avoiding the repetition of similar tasks included in both standards, and therefore, reducing the amount of effort required by the organization.

The utilization of the Security Extension is next detailed. In both cases, the organization must firstly select the ISO/IEC 27002 applicable controls, depending on the kind of organization and its main activities. As an example, we consider the implementation of the control *13.2.2 Learning from information security incidents*: "There should be mechanisms in place to enable the types, volumes, and costs of information security incidents to be quantified and monitored. The information gained from the evaluation of information security incidents should be used to identify recurring or high impact incidents. The evaluation of information security incidents may indicate the need for enhanced or additional controls to limit the frequency, damage, and cost of future occurrences or to be taken into account in the security policy review process".

In order to satisfy the security requirements related to this control, the Security Extension proposes to perform different actions in three ISO/IEC 15504-5 processes: *SUP.9 Problem resolution management* process, *MAN.5 Risk management* process and *RIN.3 Knowledge management* process.

Regarding the two first processes, *SUP.9* and *MAN.5*, their purposes should be modified or expanded as shown in section 3.4:

- *SUP.9 Problem resolution management* process must ensure that information security incidents are identified, analyzed, managed and controlled to resolution in the manner indicated by the security policy.

- *MAN.5 5 Risk Management* process must ensure that information security incidents are continuously identified, analysed, quantified, treated and monitored.

Regarding the *RIN.3 Knowledge Management* process, *RIN.3.BP4 Capture knowledge* should also be extended, as shown in section 3.2, to include the capture of information gained from the evaluation of information security incidents.

## 5 Conclusion and Further Work

In this paper, an ISO/IEC 15504 Security Extension has been presented. This Guide can be used to facilitate the implementation of ISO/IEC 27001 in software companies which are currently, or will be in the near future, involved in SPI programmes according to ISO/IEC 15504.

When a company decides to implement a particular ISO/IEC 27002 security control, the ISO/IEC Security Extension could be used to observe if the control under consideration is related to any ISO/IEC 15504-5 process. In this case, the modifications that the related process should suffer in order to cover all the security requirements proposed by the control are proposed in the Guide.

It has to be noted that, although some ISO/IEC 15504-5 processes can be easily adapted to cover some requirements of the ISO/IEC 27002 security controls, there is still a significant number of security controls that do not have any relation to ISO/IEC 15504-5 processes, and therefore, they must be implemented as indicated in the ISO/IEC 27002 standard.

Although some software companies in our environment have obtained the ISO/IEC 27001 certification after being previously involved in SPI programmes according ISO/IEC 15504, we haven't had the chance to apply the results of this research yet. However, we have initiated a new project, which includes the request for public funds to financially support this initiative.

Moreover, we are also developing a method with the guidelines for the implementation of both ISO/IEC 15504 and ISO/IEC 27001 standards reducing the amount of effort. In the case of obtaining the necessary funds to incorporate software companies as projects participants, we expect to improve our method by considering the lessons learned from its application in these software companies.

**Acknowledgments.** This research has been supported by CICYT-TIN2010-20057-C03-03 "Simulación aplicada a la gestión de equipos, procesos y servicios", Sim4Gest.

## References

1. ISO/IEC. ISO/IEC 15504-1:2004 Information Technology - Process Assessment - Part 1: Concepts and Vocabulary (2004)
2. ISO/IEC. ISO/IEC 15504-2:2003/Cor 1:2004 Software Engineering - Process Assessment - Part 2: Performing an assessment (2004)

3. ISO/IEC. ISO/IEC 27001:2005 Information technology - Security techniques - Information security management systems - Requirements (2005)
4. ISO/IEC. ISO/IEC 27002:2005 Information technology - Security techniques - Code of practice for information security management (2005)
5. Mas, A., Amengual, E.: La mejora de los procesos de software en las pequeñas y medianas empresas (pyme). In: Un nuevo modelo y su aplicación en un caso real. Revista Española de Innovación, Calidad e Ingeniería del Software (REICIS), vol. 1(2), pp. 7–29 (2005)
6. Amengual, E., Mas, A.: Software Process Improvement in Small Companies: An Experience. In: 14th European Software Process Improvement Conference, Germany, pp. 11.11–11.18 (2007)
7. Mas, A., Fluxà, B., Amengual, E.: Lessons learned from an ISO/IEC 15504 SPI Programme in a Company. In: 16th European Systems & Software process Improvement and Innovation Conference, Spain, pp. 4.13–4.18 (2009)
8. Mas, A., Amengual, E., Mesquida, A.L.: Application of ISO/IEC 15504 in Very Small Enterprises. In: Riel, A., O'Connor, R., Tichkiewitch, S., Messnarz, R. (eds.) EuroSPI 2010. CCIS, vol. 99, pp. 290–301. Springer, Heidelberg (2010)
9. Mas, A., Amengual, E.: A Method for the Implementation of a Quality Management System in Software SMEs. In: 12th International Conference on Software Quality Management, pp. 61–74. British Computer Society (2004)
10. Amengual, E., Mas, A.: A New Method of ISO/IEC TR 15504 and ISO 9001:2000 Simultaneous Application on Software SMEs. In: 3rd International SPICE Conference on Process Assessment and Improvement, The Netherlands, pp. 87–92 (2003)
11. Mesquida, A.L., Mas, A., Amengual, E.: La madurez de los servicios TI. In: Revista Española de Innovación, Calidad e Ingeniería del Software (REICIS), vol. 5(2), pp. 77–87 (2009)
12. ISO/IEC. ISO/IEC 15504-5:2006 Information technology - Process Assessment - Part 5: An exemplar Process Assessment Model (2006)
13. ISO/IEC: TR 15504-7:2008 Information technology - Process Assessment - Part 7: Assessment of organizational maturity (2008)
14. Mas, A., Mesquida, A.L., Amengual, E., Fluxà, B.: ISO/IEC 15504 best practices to facilitate ISO/IEC 27000 implementation. In: 5th International Conference on Evaluation of Novel Approaches to Software Engineering, pp. 192–198. SciTePress, Athens (2010)
15. ISO/IEC. ISO/IEC PDTR 15504-10 Information technology - Software process assessment - Part 10: Safety Extensions

# Verification & Validation in Medi SPICE

M.S. Sivakumar, Valentine Casey, Fergal McCaffery, and Gerry Coleman

Regulated Software Research Group,
Dundalk Institute of Technology & Lero, Ireland
{smadh09@studentmail,Val.Casey,Fergal.McCaffery,
Gerry.Coleman}@dkit.ie

**Abstract.** Effective verification and validation are central to medical device software development and are essential for regulatory approval. Although guidance is available in multiple standards in the medical device software domain, it is difficult for the manufacturer to implement as there is no consolidated view of this information. Likewise, the standards and guidance documents do not consider process improvement initiatives. This paper assists in relation to both these aspects and introduces the development of processes for verification and validation in the medical device domain.

**Keywords:** Medical device standards, Medical device software verification and validation, Medical device software process assessment and improvement.

## 1 Introduction

Verification and Validation (V&V) activities are important activities in the software development lifecycle and consume up to 50% of project development time [1], [2] and up to 50% of the total cost [3]. While both V&V play a key role in software development, there is a level of ambiguity in the use of these terms. This is evident from the difference in definition of these terms in the literatures [4], [5] and [6].

When developing safety-critical software it is imperative to have software development practices which incorporate effective V&V activities. In this context V&V are addressed by numerous standards for both generic and safety-critical software development which include specific medical device standards.

The National Institute of Standards and Technology performed a study, indicating that software defects cost the U.S. economy in the region of $59.5 billion a year [7]. The study also indicates that better testing could detect and remove defects early in the development process and reduce the cost by more than a third [7]. However, there are challenges in the implementation of V&V in the context of general software development and these challenges are even greater in safety-critical domains. The requirements put forth by the regulatory bodies stress the need for supporting documentation and it can be challenging to satisfy these regulatory requirements and meet the pressures of the market at the same time.

## 2 V&V in Generic Software Development

Two important reference models which are widely used in the context of software process improvement are the Capability Maturity Model® Integrated (CMMI®) [8]

and ISO/IEC 15504-5:2006 [9]. When considering software V&V it is of value to consider both.

CMMI® recommends a lifecycle approach for V&V activities. It defines verification as "Confirmation that work products properly reflect the requirements specified for them". In other words, verification ensures that "you built it right" and validation is "Confirmation that the product, as provided (or as it will be provided), will fulfill its intended use". In this context, validation ensures that 'you built the right thing'. The V&V processes are part of the engineering processes category and both are level 3 process areas in the staged model. The model also provides guidance in terms of examples of methods such as peer reviews; statement coverage testing and branch coverage testing that could be performed. The validation process area incrementally validates products against the customer's needs. Validation may be performed in the operational environment or simulated operational environment. Coordination with the customer in relation to validation requirements is an important element of this process area. The scope of the validation process area includes validation of products, product components, selected intermediate work products, and processes. These validated elements may often require re-verification and re-validation.

In ISO/IEC 15504-5:2006 V&V are two distinct processes and are part of the supporting lifecycle process group. Both of these processes are based on the respective lifecycle processes in ISO/IEC 12207 AMD1 [10]. In ISO/IEC 15504-5, the purpose of the verification process is to confirm that each software work product and/or service of a process or project properly reflects the specified requirements. The tasks pertaining to verification include: development of a verification strategy, development of criteria for verification, performing the activity of verification, determination of actions based on verification results and making the results available to the stakeholders.

Industry experience indicates that V&V activities typically consume about 30-50% of development budgets [11]. CMMI® and ISO/IEC15504-5 are not prescriptive when it comes to methods and tools to be used for V&V. Rather it is left to the discretion of the user to select and apply methods. Though CMMI®, considers validation and verification, it is still rather modest in its focus on these areas compared to other elements of the development processes [12].

## 3 V&V for Safety-Critical Software Development

Software can be a critical element of complex, potentially dangerous products such as weapons systems, aerospace systems and medical devices. These are critical because failure can result in loss of life, significant environmental damage, and major financial loss [13]. It has also been found that there is a relationship between the increasing occurrence of system accidents and the increasing usage of software [14]. In these circumstances these products are required to meet a very high-level of reliability, security, and performance. Therefore, ensuring that such systems meet their predefined requirements and that they perform as expected is an essential and often challenging issue [15].

Within the safety-critical software arena, different standards/certifications are available for different industries. These include the MIL-STD-498 [16] for military applications, DO-178B [23] for Aerospace, and Automotive SPICE and ISO 26262 [17] in the Automotive industry. IEC 60880 [18] describes the European standards for the certification of nuclear power generating software. IEC 61508 [19] describes a general-purpose hierarchy of safety-critical development methodologies that has been applied to a variety of domains ranging from medical instrumentation to electronic switching of passenger railways. Though these standards address V&V in sufficient detail, their role is not to address process improvement. In addition, there are some [20] who consider that a CMMI® V&V assessment inadequate when dealing with safety-critical software, and they propose a new framework for V&V assessment, focused on safety-criticality. This framework is defined through integrating safety standards with the V&V process areas of the CMMI® and the ISO 9001 standard [21].

The following are some of the attributes of safety standards: (1) Product versus process (2) Safety management agents (3) Risk assessment (4) Integrity levels (5) Design safety and (6) Assurance techniques [22]. Based on these criteria, we decided to use DO-178B and the Automotive SPICE as part of our research for developing V&V processes for the medical device software domain. Automotive SPICE has been derived from ISO/IEC 15504-5. This was of particular relevance as it was developed for a safety-critical domain to facilitate process assessment and improvement.

Therefore, existing software process reference models need to be adapted and extended to meet the specific requirements of medical device software development which is safety-critical in nature.

## 4 Research Approach and Outcomes

The research involved a number of stages:

1. The V&V processes were reviewed in detail and consideration was given to how they were addressed by generic software development standards and process improvement reference models, which included ISO/IEC 12207, ISO/IEC 15504-5 and CMMI®.
2. A literature review of V&V was performed in terms of safety-critical software development. This included a review of the V&V processes addressed by safety-critical software development standards such as DO-178B and Automotive SPICE [24].
3. A literature review and analysis was also performed in relation to medical device software V&V. This incorporated the Food and Drug Administration (FDA) guidelines for Software Validation (FDA GPSV) [25], the Medical Device Directive (MDD) 1993/42/EEC [26] and amendment 2007/47/EC [27], ISO/IEC 62304 [28], ISO/IEC 13485 [29] and ISO/IEC 14971 [30].
4. Based on this analysis we defined a set of processes for V&V for medical device software development. The processes were assigned a Process ID, Process Name, Process Purpose, Process Outcomes and a set of Specific Practices. Guidance in the implementation of these specific practices is provided through a set of sub-practices and notes. These processes were developed as part of the Medi SPICE [41].

## 4.1 Regulatory Nature of the Medical Device Domain

Studies in the medical device industry [31], [32], point to the fact that software is one of the most critical factors for cutting edge products and the role software plays is predicted to continue to increase [32]. It is also expected that, by 2015, the research and development investment in software in this area will increase from 25% of the overall budget in 2002, to 33% [32].

However, as the role of software in the medical device domain increases, so do the number of failures which arise due to software defects. An analysis of medical device recalls by the FDA in 1996 [33] found that software was increasingly responsible for product recalls. This continues to be the case and in the period the $1^{st}$ January 2010 to $1^{st}$ January 2011 the FDA recorded 80 medical device recalls and state software as the cause [46]. A German survey on medical device recalls indicated that software was the top cause for risks related to construction and design defects of medical device products. This analysis, from June 2006, showed that 21% of the medical device design failures were caused by software defects. This was an increasing trend, as the figures from November 2005 showed software was responsible for 17% of construction and design defects [34].

Due to the safety-critical nature of medical devices, medical device companies who wish to sell their products must comply with the regulatory requirements of the respective countries where they plan to market them. Medical devices can only be sold in the US if they comply with the FDA regulations [35], whereby a quality system needs to be in place that complies with the FDA Regulations 21 CFR Part 820, Quality System Regulation (QSR) [36]. In order to sell devices the manufacturer not only has to prove safety and effectiveness, but also has to demonstrate that the design and development of the device including the software complies with the FDA regulations. The "Guidance for the Content of Premarket Submissions for Software Contained in Medical Devices" document [37] details these requirements. Though the regulatory bodies, such as the FDA provide guidance documents, they do not dictate that a particular method must be used [38]. The quality system process itself is designed by the medical device manufactures and the quality system process needs to ensure that the manufacturer is designing and building a quality product. The difficult part is that the manufacturer has to provide evidence to the FDA inspectors that the correct processes have been followed [35].

In order to achieve standardization of expectations and for better guidance for implementation by manufacturers, the FDA Center for Devices and Radiological Health (CDRH) has published guidance documents which include risk-based activities to be performed during software validation [25], pre-market submission [37] and when using off-the-shelf software in a medical device [39]. Although the CDRH guidance documents provide information on which software activities should be performed, they do not enforce any specific method for performing these activities. The result is that the medical device manufacturers could fail to comply with the expected requirements.

Within the medical device industry a decision was initially made to recognize ISO/IEC 12207:1995 (a general software engineering lifecycle process standard) as being suitable for general medical device software development [10]. However, upon careful examination of ISO/IEC 12207, the Association for the Advancement of

Medical Instrumentation (AAMI) software committee decided it was necessary to create a new standard specifically for medical device software development. The AAMI used ISO/IEC 12207:1995 as the foundation for their new standard "AAMI SW68, Medical device software – Software lifecycle processes" [38]. In 2006, a new standard AAMI/IEC 62304 [28] was released that was based on the AAMI SW68 standard.

In order to sell medical devices within the Europe Union (EU) the CE mark is required. To achieve the CE mark compliance is required with the Medical Device Directive (MDD) (1993/42/EEC) and amendment MDD (2007/47/EC), In-Vitro Diagnostic Directive (IVDD) [44] and the Active Implantable Medical Device Directive (AIMDD) [45] depending on the type of medical device being submitted. As stated in the latest amendment to the MDD, Section 1 (g) of MDD (2007/47/EC) "For devices which incorporate software or which are medical software in themselves, the software must be validated according to the state of the art taking into account the principles of development lifecycle, risk management, validation and verification". "State of the Art" is used here to mean what is generally accepted as good practice. Since this requirement was introduced, developers must now validate the software be it integrated or standalone, regardless of device class. IEC 62304 and its aligned standards are often seen as a good place to start when validating software.

Whilst these standards are generally accepted and are harmonised under the MDD they do contain omissions which make them difficult to apply to standalone software as an active medical device. As we had observed from our research this is exemplified by IEC 62304 where there is no provision within the standard to validate the system elements of standalone software.

While there are numerous standards in the medical device domain they are oriented towards achieving regulatory compliance. As a result the focus of medical device software development is compliance rather than process improvement. To address this Medi SPICE is being developed. The objective of Medi SPICE is to provide a process assessment and improvement model which is domain specific to medical device software development and incorporates regulatory compliance. Medi SPICE will also enable the harmonization of different standards in the medical device software development domain, thus bringing best practices available in multiple standards into a single framework which will aid manufacturers in the implementation of their requirements as well as in their process improvement exercise. The results of a Medi SPICE assessment may be used to indicate the state of a medical device suppliers software practices in relation to the regulatory requirements of the industry, and identify areas for process improvement. The results of these assessments may also be used as a criterion for supplier selection. The authors believe that, with the publication of the Medi SPICE more specific guidance will be available for the basis of process design and assessment in the medical device industry [41].

### 4.2 V&V in Medi SPICE

Based on our research which comprised of an extensive literature review and comparative analysis of standards in the medical device and other safety-critical domains, we arrived at the following findings, which were incorporated into the definition of the processes related to V&V in Medi SPICE:

a. From the literature review and comparison across other standards and models, it became clear that the terms V&V are frequently used interchangeably. The FDA guidelines distinguish between verification and validation. Though the FDA is clear on the definition part, sections 4, 5 and 6 of the FDA GPSV, which deals with operational activities, still use the term validation only and no reference is made to verification. Guidance on differentiating between V&V activities with respect to the different engineering activities/work products should be in place. The amendment 2007/47/EC to the MDD stresses the importance validation plays and the need for state of the art validation and verification
b. Verification is not addressed as a separate process in IEC 62304 and verification practices are integrated into other engineering processes. Validation is considered a systems level process and outside the scope of IEC 62304 even when the system consists entirely of software.
c. Automotive SPICE has V&V criteria and V&V records as outputs in its processes. The ISO/IEC 15504-5 does not go to this level of detail.
d. The IEEE Standards for Software V&V state that classical Independent Verification & Validation (IV&V) is generally required for the development of software systems deemed "critical", i.e., those which can result in loss of life, mission or significant social or financial loss [42]. Independence is an important factor addressed by DO-178B. The degree of independence is also addressed in ISO/IEC 15504-5 and Automotive SPICE. The FDA GPSV Sec 4.9 does address independence and leaves it to the discretion of device manufacturers as to how this is to be achieved. Independence is not addressed as part of IEC 62304 and it assumes that it is taken care of by/through ISO 13485. Therefore, in Medi SPICE we placed a particular focus on the clarity of independence in the verification and validation processes.
e. Software developed for medical devices concerns itself with obtaining regulatory approval as opposed to improving processes to achieve more efficient software development [40]. Models like CMMI® and ISO/IEC 15504-5 have separate process areas for verification and validation. A separate process area for critical processes like V&V enable organizations to understand their strengths and weaknesses in a detailed manner and can provide help when embarking on process improvement initiatives.

From our analysis of the relevant literature regarding V&V and through the comparative analysis of process improvement models and standards, our goal was to determine best practice in this area and to facilitate process improvement. Our objective was also to satisfy the requirements of the relevant medical device standards which include the FDA GPSV, MDD, ISO/IEC 13485, IEC 62304, IEC TR 80002-1 [43] and ISO14971. Having established these elements, it was imperative we incorporate them into Medi SPICE. To achieve this we developed the following Medi SPICE processes with a particular emphasis on verification and validation:

1. Software Construction
2. Software Integration
3. Software Testing
4. Verification
5. Validation

As discussed in Section 4, our objective was to incorporate the relevant standards and the most effective elements of process improvement models into a common framework specifically designed for the medical device software domain, Medi SPICE.

Unlike ISO/IEC 15504-5, where there is no requirement for the classification of outcomes and processes based on safety, this was an important element which had to be included in Medi SPICE. We therefore utilized the classification schema provided by IEC 62304, which is used to associate the outcomes and specific practices with the safety level of the software for which the practices are applicable. While we based the practices on the ISO/IEC15504-5, our analysis of standards in similar safety-critical domains highlighted that it would be beneficial to use Automotive SPICE as our reference, as it is has been developed to meet the specific requirements of safety-critical software development. As a result of the findings from our research, we have included V&V as separate process areas in Medi SPICE. The validation process includes many of the recommendations that were produced as part of this research.

Risk management is an integral part of medical device software development. In this context the relevant standards for medical device development are ISO 14971 and IEC 62304. IEC/TR 80002-1 provides specific guidance as to how these two standards can be combined to address risk with regard to medical device software development. The requirements of V&V as required by these standards have been addressed in the five processes listed above.

### 4.3 Summary of V&V Related Processes in Medi SPICE

For the purpose of this paper, we use the Software Testing (ENG.8) process in Medi SPICE as an example. This process relates to the IEC 62304 Software System Testing activity which has five tasks. As an outcome of our analysis, specific practices (1 to 10) were defined for the Software Testing Process. The practices and how they map to relevant medical device standards are illustrated in the Table 1:

Against the five tasks that the IEC 62304 provides, the Medi SPICE Software Testing Process has nine specific practices and one sub practice. In line with the good practice of ensuring traceability at each engineering activity level as observed in ISO/IEC 15504-5, Medi SPICE also focuses on using traceability in each engineering activity as it is very important from a verification perspective. In addition, to the specific practices a single sub-practice, additional implementation guidance is provided through 10 notes in the Software Testing Process. It may be noted from Table 1 that a specific practice – Conduct risk control activities has been added as part of the model. We are thus providing guidance through Medi SPICE on risk management activities in line the ISO 14971, which requires verification of the implementation of risk control, as well as verification of the reduction of risk through adopting risk control mechanisms.

**Table 1.** Mapping ISO/IEC 62304 Tasks to Medi SPICE practices

| IEC 62304 Reference | Sub Task / Clause | Medi SPICE Reference | Medi SPICE Reference |
|---|---|---|---|
| 5.7.1 | Establish tests for software requirements | ENG.8.SP1 | Develop software test strategy |
| | | ENG.8.SP1.1 | Establish a set of tests |
| | | ENG.8.SP2 | Develop test specification for software test |
| | | ENG.8.SP4 | Test the integrated software |
| 5.7.2 | Use software problem resolution process | ENG.8.SP5 | Record the anomalies |
| 5.7.3 | Retest after changes | ENG.8.SP9 | Develop regression test strategy and perform regression testing |
| 5.7.4 | Verify software system testing | ENG.8.SP3 | Verify test specification for software test |
| | | ENG.8.SP7 | Verify software testing |
| 5.7.5 | Software system test record contents | ENG.8.SP6 | Record results of software test |
| | | ENG.8 SP8 | Ensure consistency and bilateral traceability |
| | | ENG.8.SP10 | Conduct risk management activities |

Table 2, outlines how we have addressed some of the typical software testing tasks, with reference to the FDA GPSV guidance document.

**Table 2.** Mapping FDA Typical tasks to Medi SPICE

| FDA Typical Tasks | Medi SPICE Reference |
|---|---|
| Test Planning | Software Construction |
| Functional test case identification | Software Construction |
| Traceability analysis | Software Construction, Software Integration, Software Testing |
| Unit (module) test execution | Software Construction |
| Integration test execution | Software Integration |
| Functional test execution | Software Integration |
| System test execution | Software Testing |
| Error evaluation/resolution | Software Testing |
| Final Test Report | Software Testing |

The requirements of FDA GPSV are directly addressed by Medi SPICE as can be observed from the mapping. Further, it needs to be noted that the task of Acceptance test execution is not addressed by Medi SPICE as part of software engineering processes. This is in line with the Automotive SPICE as well as ISO/IEC 15504-5, where acceptance testing is part of the acquisition group of processes.

## 5 Conclusions and Future Work

Further to the definition of a set of process areas and the associated practices related to V&V, the processes should be piloted in organizations within the medical device software development industry. Based on the results observed, the processes should be evaluated and continuously improved based upon feedback from the medical device software development industry.

Additionally, as V&V absorbs a significant amount of project time, further research will be performed on practices, which could bring in reduction in cycle time for V&V activities but without compromising quality and safety features of the products being developed.

Globalization of software development has led to distributed teams working on the same product in different locations. Understanding the challenges in globally distributed V&V in the context of medical device software development and what additional practices could aid practitioners in such cases. These practices could then become notes or sub-practices in further versions of Medi SPICE.

As medical device manufacturers outsource their medical device software development, it would be worthwhile to examine: (a) what could be drivers in outsourcing V&V activities to a third party. (b) if outsourcing of medical device software development is performed, will V&V still be done internally? (c) What risks should be considered and practices should be included in a reference model for medical device software development from a V&V perspective for 3rd party software (COTS) or where certain activities are outsourced?

## Acknowledgments

This research is supported by the Science Foundation Ireland (SFI) Stokes Lectureship Programme, grant number 07/SK/I1299, the SFI Principal Investigator Programme, grant number 08/IN.1/I2030 (the funding of this project was awarded by Science Foundation Ireland under a co-funding initiative by the Irish Government and European Regional Development Fund), and supported in part by Lero - the Irish Software Engineering Research Centre (http://www.lero.ie) grant 03/CE2/I303_1.

## References

1. Bernard, E., Legeard, B., Luck, X., Peureux, F.: Generation of test sequences from formal specifications: GSM 11-11 standard case study. Software: Practice and Experience 34(10), 915–948 (2004)
2. Rayadurgam, S., Heimdahl, M.P.E.: Test-sequence generation from formal requirement models. In: Proc. Sixth IEEE International Symposium on High Assurance Systems Engineering, pp. 23–31 (2001)
3. Myers, G., Sandler, C., Badgett, T., Thomas, T.: The Art of Software Testing, 2nd edn. Wiley, Chichester (2004)

4. Wallace, D.R., Fujii, R.U.: Software verification and validation: an overview. IEEE Software 6(3), 10–17 (1989)
5. Berling, T., Thelin, T.: An industrial case study of the verification and validation activities. In: Proceedings of Ninth International, Software Metrics Symposium, September 3–5, pp. 226–238 (2003)
6. Arthur, J.D., Groner, M.K., Hayhurst, K.J., Holloway, C.M.: Evaluating the effectiveness of independent verification and validation. Computer 32(10), 79–83 (1999)
7. Tassy, G.: The Economic Impacts of Inadequate Infrastructure for Software Testing. National Institute of Standards and Technology (NIST), report 02-3 (May 2002)
8. CMMI Product Team, Capability Maturity Model® Integration for Development Version 1.2. Software Engineering Institute (2006)
9. ISO/IEC 15504-5:2006, Information technology — Process Assessment — Part 5: An Exemplar Process Assessment Model. ISO, Geneva, Switzerland (2006)
10. ISO/IEC 12207:1995/Amd.1, Information Technology — Software life Cycle Processes Amendment 1. ISO, Geneva, Switzerland (2002)
11. Kit, E.: Software Testing in the Real World. Addison-Wesley, London (1995)
12. Jacobs, J.C., Trienekens, J.J.M.: Improving verification and validation in hardware/software environments. In: Introduction to the workshop. Proceedings of 10th International Workshop on Software Technology and Engineering Practice, STEP 2002, October 6-8, pp. 121- 122 (2002)
13. Why Software Requirements Traceability Remains a Challenge - Cross Talk, Issue (July/August 2009), http://www.stsc.hill.af.mil/crosstalk/2009/07/0907KannenbergSaiedian.html
14. Leveson, N.G.: Paper on System Safety in Computer-Controlled Automotive Systems. Massachusetts Institute of Technology, http://sunnyday.mit.edu/papers/sae.pdf
15. Alawneh, L., Debbabi, M., Hassaine, F., Jarraya, Y., Soeanu, A.: A unified approach for verification and validation of systems and software engineering models. In: 13th Annual IEEE International Symposium and Workshop on Engineering of Computer Based Systems, ECBS 2006, March 27-30 pp. 409–418, (2006)
16. MIL-STD-498. Military Standard - Software Development And Documentation. US Department of Defense, USA, December 5 (1994)
17. ISO/DIS 26262 Road vehicles - Functional safety. ISO, Geneva, Switzerland
18. IEC 60880:2006, Nuclear power plants - Instrumentation and control systems important to safety - Software aspects for computer-based systems performing category A functions. IEC, Geneva, Switzerland (2006)
19. IEC/TR 61508:2005, Functional safety of electrical/electronic/ programmable electronic safety related systems. BSI, London (2005)
20. Kyung, A.Y., Seung-Hun, P., Doo-Hwan, B., Hoon-Seon, C., Jae-Cheon, J.: A Framework for the V&V Capability Assessment Focused on the Safety-Criticality. In: 13th IEEE International Workshop, Proc. Software Technology and Engineering Practice, pp. 17–24 (2005)
21. ISO 9001:2000 - Quality management systems - Requirements (2000)
22. Eastaughffe, A., Cant, A., Ozols, M.A.: A Framework for Assessing Standards for Safety-critical Computer-Based Systems. In: Fourth IEEE International Symposium and Forum on Software Engineering Standards. ISESS, p. 33 (1999)
23. RTCA, RTCA DO-178B, Software Considerations in Airborne Systems and Equipment Certification (1992)
24. Automotive SPICE Process Assessment. SIG, August 21 (2005)
25. US FDA Center for Devices and Radiological Health. General Principles of Software Validation; Final Guidance for Industry and FDA Staff. CDRH, Rockville (2002)

26. European Council, Council Directive 93/42/EEC Concerning Medical Devices. Official Journal of The European Communities, Luxembourg (1993)
27. European Council, Council Directive 2007/47/EC (Amendment). Official Journal of The European Union, Luxembourg (2007)
28. ANSI/AAMI/IEC 62304:2006, Medical device software—Software life cycle processes. AAMI, Arlington (2006)
29. ISO 13485:2003. Medical devices — Quality management systems — Requirements for regulatory purposes, 2nd edn. ISO, Geneva, Switzerland (2003)
30. ISO 14971:2007. Medical Devices — Application of risk management to medical devices, 2nd edn. ISO, Geneva (2007)
31. Future Trends in Medical Device Innovation. Advanced Medical Technology Association, AdvaMed (2004),
    http://www.advamed.org/MemberPortal/About/NewsRoom/MediaKits/futuretrendsinmedicaldeviceinnovaton.htm
32. ITEA 2 Blue Book (September 2005),
    http://www.itea2.org/itea_2_blue_book
33. Wallace, D.R., Richard Kuhn, D.: Failure Modes in Medical Device Software: An analysis of 15 years of recall data. International Journal of Reliability, Quality and Safety
34. Feldmann, R.L., Shull, F., Denger, C., Host, M., Lindholm, C.: A Survey of Software Engineering Techniques in Medical Device Development. In: Joint Workshop on High Confidence Medical Devices, Software, and Systems and Medical Device Plug-and-Play Interoperability, HCMDSS-MDPnP, June 25-27, pp.46-54 (2007)
35. Lin. W., Fan, X.: Software Development Practice for FDA-Compliant Medical Devices. In: International Joint Conference on Computational Sciences and Optimization, CSO 2009, April 24-26, vol.2, pp. 388–390 (2009)
36. US FDA/CDRH, 21CFR820, Quality System Regulation (2007)
37. US FDA/CDRH, Guidance for the Content of Premarket Submissions for Software Contained in Medical Devices (2005)
38. Burton, J., McCaffery, F., Richardson, I.: A risk management capability model for use in medical device companies. In: International Workshop on Software quality (WoSQ 2006), Shanghai, China, pp. 3–8. ACM, New York (2006)
39. US FDA Center for Devices and Radiological Health. Off-The-Shelf Software Use in Medical Devices; Guidance for Industry, medical device Reviewers and Compliance. CDRH, Rockville (1999)
40. Eagles, S., Murray, J.: Medical Device Software Standards: Vision and Status, Medical Device and Diagnostic Industry (May 2001),
    http://www.mddionline.com/archive/all/1969?page=475
41. McCaffery, F., Dorling, A., Casey, V.: Medi SPICE: An Update. In: The 10th International Spice Conference, SPICE 2010, Pisa, Italy May 18–20 (2010)
42. IEEE Standards for Software Verification and Validation, p. 58. IEEE, Los Alamitos (1998)
43. IEC/TR 80002-1:2009. Medical device software Part 1: Guidance on the application of ISO 14971 to medical device software. BSI, London (2009)
44. European Council, Council Directive 98/79/EC On in vitro diagnostic medical devices. Official Journal of The European Communities, Luxembourg (1998)
45. European Council, Council Directive 90/385/EEC On the approximation of the laws of the Members States relating to active implantable medical devices. Official Journal of The European Communities, Luxembourg (1990)
46. Medical & Radiation Emitting Device Recalls, FDA (2011),
    http://www.accessdata.fda.gov/scripts/cdrh/cfdocs/cfres/res.cfm
    (accessed February 10, 2011)

# Medical Device Software Development - A Perspective from a Lean Manufacturing Plant

Oisín Cawley, Ita Richardson, and Xiaofeng Wang

Lero - the Irish Software Engineering Research Centre,
University of Limerick
Ireland
{oisin.cawley,ita.richardson,xiaofeng.wang}@lero.ie

**Abstract.** Developing software for the manufacture of medical devices is a sensitive operation from many perspectives, such as: safety and regulatory compliance. Medical Device companies are required to have a well defined development process in place, which includes software development, and be able to demonstrate that they have followed it through the complete life-cycle of the device. With the increasing complexity of Medical Devices, and more detailed software development regulations among some of the influencing factors, we take a look at how some of these factors have impacted the software development process within a medical device manufacturing plant. We find that tying down your process across the board can have unwanted consequences. As process flexibility is required, we have investigated the usefulness of Lean Software Development.

**Keywords:** Software Development Process, Medical Device, Regulated Environment, Process Improvement, Lean Software Development.

## 1 Introduction

As would be expected, the concern for human safety takes precedence when it comes to the development of a Medical Device (MD). To ensure the lowest level of risk to safety, various regulatory controls have been established governing the development process. For example within the European Union, the Medical Device Directive [1], [2] define a set of harmonised rules relating to the safety and performance of MDs. One particular aspect which has gained increased attention from a regulatory perspective is the software development life-cycle (SDLC). The software component of a MD is playing an increasingly important role in the construction and operation of MDs, and is becoming more and more complex. This has been reflected in the relatively recent addition to the definition of a MD, for example by the U.S. Food and Drugs Administration (FDA) [3], to include software in its own right as a possible MD.

One thing which is worth noting is that MD software covers embedded-software and also software involved in the manufacturing of the device. It is within this context that we examine the software development process of an Irish based MD manufacturing plant to see how they configure their internal processes, and what the influencing factors are in achieving regulatory compliance.

## 1.1 Software for the Medical Device Industry

Medical devices can be defined as being safety-critical. Other domains which fall into this category are the Aviation, Automobile, Railway, and Nuclear among others. As mentioned above, the obvious concern is around safety, and the aim is to minimise the risk of injury to humans to an acceptable level. Each of these safety-critical domains has developed and published standards and guidelines to help achieve the safest possible end-product. For example: the DO-178B guidelines for airborne systems and equipment [4], and the U.S. code of federal regulations title 21 part 820 governing the quality system regulations for medical device manufacturers [5].

Since software is increasingly becoming an integral part of a MD, we have seen an increase in the number of injuries caused to patients which have been directly attributed to the software component. FDA analysis of 3,140 medical device recalls between 1992 and 1998 found that 242 (over 7%) were attributable to software [6]. Significantly, of the recalls in 2007 of what the FDA classify as life-threatening, 23 of them involved faulty software [7]. As a consequence, regulatory controls are continuously being reviewed and adapted to this ever advancing technological landscape. In 2006 an international standard (ANSI/AAMI/IEC 62304:2006) was published which governs the MD SDLC processes [8]. Now widely adopted, IEC 62304 establishes a common framework for medical device life-cycle process by describing a set of processes, activities, and tasks that are required within a MD SDLC.

However, reading the standards can lead to thinking that a waterfall-type software development methodology is what will best meet the requirements. This is in fact not the case, and Annex B of the IEC 62304 standard specifically clarifies that: "This standard does not require a particular SOFTWARE DEVELOPMENT LIFE CYCLE MODEL". This allows companies to employ whatever methodology they prefer, for example, Incremental or Evolutionary. Typically MD companies employ a traditional SDLC model (waterfall or V), but lately, more focus is being given to examining how these companies can improve their SDLC processes, for example by employing a more iterative development methodology [9], [10].

## 1.2 Lean Software Development

The concept of Lean Software Development can be thought of as the merging of the principles of Lean Manufacturing [11], [12], with software development practices. Lean's primary focus is on the identification and elimination of waste from the process. Waste being defined as "any human activity which absorbs resources but creates no *value*" [13]. So lean thinking "is lean because it provides a way to do more and more with less and less – less human effort, less equipment, less time, and less space – while coming closer and closer to providing customers with exactly what they want" [13].

Eliminating waste, when translated into software engineering terms, can mean the elimination of defects (bugs) in the code. This may seem an obvious goal of developing software, but the creation of formal mechanisms to achieve this began to show the power of doing this in a systematic fashion. This is one of the cornerstones of what Lean Software Development is founded on, finding and fixing defects early in the development process. Many of the Agile software development practices [14], [15] can be seen as supportive of a lean philosophy (Fig. 1). For example, the agile practice of test driven development (TDD) [16] in order to find defects early by continuous testing and thus reducing the cost of rework later. Many more such practices have been mapped by [17] and [18].

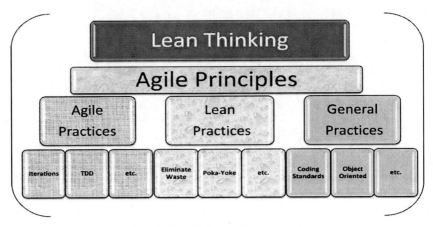

**Fig. 1.** Lean Software Development

Of interest therefore is how such 'lean practices' can be utilised in the MD domain, something the industry is also looking at [10]. With a focus on an iterative approach to software delivery, and favouring less rather than more documentation [19], companies can shy away from such practices out of fear of a costly non-compliance outcome. Slowly, however, we are seeing more and more reports from companies who are trialling lean approaches in this domain [20], [21], [9].

## 2 Research Approach

Following on from a systematic literature review of software development within safety-critical regulated environments [22], we were interested in investigating further how the various regulations and other influencing factors affect the software development process within the MD domain, and how a lean perspective could be beneficial. Our approach to this was to undertake a case study within a MD company, using [23], [24] as guides. Taking an interpretative approach, we requested one-to-one interviews with a cross section of the organisation involved in projects which had a software development component. Eight onsite interviews were performed lasting between sixty and ninety minutes each. The interviewees all had relevant firsthand experience of the complete product development process and the governing policies and procedures. They included senior software developers, a senior quality engineer, a process engineer, and project leaders. Their work experience ranged from 7 to 18 years, and from 3 to 9 years within a MD context.

With the permission of the interviewees, the interviews were recorded and later transcribed and analysed using qualitative data analysis techniques such as open and axial coding as described by [25]. Other artefacts were also gathered while on site, such as documents describing internal processes and procedures, organisational charts, project metrics, corporate policies, standards, presentations, and email correspondence. Together with on site observations, all these artefacts were used within the case study to gain a more holistic view of the working environment.

### 2.1 The Company

MedTech (a pseudonym) is a large US medical device company with manufacturing facilities located in the United Sates and Ireland. Within the particular plant we investigated, the MDs do not have any embedded software, but a large effort is required in developing and maintaining the automation software necessary for manufacturing the devices. The plant performs a combination of research and development as well as commercial MD manufacturing.

Like most businesses, the current global economic conditions have also taken their toll on MedTech. They went through a process of workforce reduction in recent years, and during our research we noted how this reduction has affected the way the employees work. As a consequence the daily endeavours of how they comply with the regulations, has been brought into the spotlight, something we will discuss below.

As a committed Lean organisation, MedTech maintains quite an impressive visual display of their values, lean initiatives and achievements. What was interesting from the interviews therefore were the responses to questions probing software development process improvement from a lean point of view.

## 3 Research Findings

From our case study analysis, we show the key components which have shaped the development process within MedTech (Fig. 2).

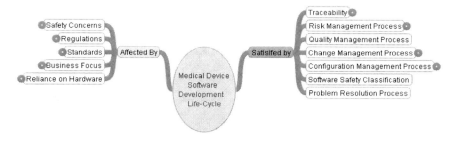

**Fig. 2.** Overview of key influences on the SDLC within MedTech

Due to space limitations we expand here on just two of the important factors which emerged from the case study, namely: Regulations and Business Focus. These factors hugely influenced how the SDLC evolved to what it is today, and continue to exert pressure on various processes within the organisation, but particularly the product development process, of which the SDLC is a key sub-process.

## 3.1 Regulations

Ensuring that the SDLC meets the requirements of the regulations typically means that internal processes are defined and documented and have been mapped to a relevant standard. For example, an internal risk management process may be mapped to the international standard ISO 14971:2007 (international standard for the application of risk management to medical devices). However, these standards are open to interpretation and even with the aid of guidance documents such as those published by the FDA's CDRH (Centre for Devices and Radiological Health), and also due to the different classifications of MDs, it can be difficult to know exactly what the auditors expect. As a result there may be a tendency to overdo it in terms of the process, in a 'just to be sure' approach.

MedTec's growth was partly due to the acquisition and amalgamation of smaller companies. Each of these had their own internal processes, and little attention was given to the actual software development process. As stated during an interview: *"The development bit in the middle there was a bit of a black art"*. However following an Audit in 2005, a deficiency was found, in that they were unable to show traceability back to a user requirement for a particular controlling software parameter. The serious nature of this spurred them into a corporate initiative to revamp their processes. A corporate wide SDLC was defined governing all aspects of software development and which mapped to the relevant regulations.

This corporate umbrella-policy was designed to cover all relevant activities: system upgrades, new product design, and production issue resolution. It was therefore necessarily high level, and consequently another level of granularity was required to govern at a site level. The site level SDLC process they implemented mapped to the corporate policy and was therefore considered to be in compliance with the regulations. The various teams trust that these policies are regulatory compliant. When asked specifically about IEC 62304 and how it affects their processes, a response was: *"We've got a regulatory group that assess standards against corporate policy...So I assume that activity would have happened up there"*.

But what is evident from the case study is that, three or so years on, there is widespread discontent with the process. This has been acknowledged at a corporate level, and they have kicked off an initiative to review it. The huge amounts of documentation, the number and level of approvals, the time required, the cost, were the types of issues the interviewees had with the process. The following quote summarises the mood: *"You can do all this [the process] and deliver [bad quality] to your customer ... which is why it's seen as a failure in the overall organisation. Project managers and R&D complain about it, engineering managers complain about it, nobody understands it, it's just a noose, it's just a pain really*".

What we have observed is that, by means of an initiative to improve the software development process, some very undesirable side effects have occurred. One of the reasons for this is that the process is indiscriminate of the focus of the activity being carried out. Within the R&D projects for example, the nature of the work is that they do not know up front exactly what it is they want the software to do. A rough idea of the requirements is known but many of the intricacies, for example, what margin of error is required, will only be known once the software and equipment have been prepared and trialled. So while the process calls for full requirements disclosure up front, followed by a lengthy change control process for any subsequent changes, this does not lend itself well to an iterative-type product development process typical of most R&D activities. Of one project, the following was said of the software process, "*... [it would] add 4 months onto the schedule of delivering the equipment to the customer*". Such considerable overhead can make people look for short cuts, and that can't be seen as a good thing.

### 3.2 Business Focus

Breaking down the components of the business focus category we see the main influencers as illustrated in Fig. 3:

**Fig. 3.** Component elements of the Business Focus factor within MedTech

Again due to space restrictions, we describe two of these, Cost and Time-To-Market which were the highest ranked in terms of the coding process, and briefly discuss how each one affects the software development process.

### 3.2.1 Cost

Developing Medical devices is an expensive business, both in terms of the opportunity cost (time to market, which we present in section 4.2.2) and the cash burning from simply running the whole process. In a recent U.S. economic report [26], the premium paid to employees in the Medical Technology Industry (MTI) is highlighted: "One of the outstanding characteristics of the medical technology industry is the strong pay scale ... almost a 40 percent premium for jobs".

MedTech also went through a process of headcount reduction to reduce costs. As a result some process improvement initiatives have suffered by being either cancelled or postponed. Another effect has been that product managers are less willing to pay for software development unless it is really necessary. Because

MedTech operate an internal cross-charging process, managers are less likely to proceed with software or equipment enhancements once they see the project estimate. Thus the effect on the developers is that they spend a significant amount of time producing estimates for software development which might never proceed. From a lean perspective this type of task-switching introduces waste into the process and should be minimised: "Every time software developers switch between tasks, a significant switching time is incurred as they get their thoughts gathered and get into the flow of the new task" [17].

MedTech also utilises contract resources on an as-needed basis which can be hugely expensive *"We pay contract validation resources to do all this work, to the scale of millions per year"*. However the duration of the software validation cycle can be quite difficult to estimate, and so when contract resources are employed, costs can very quickly accumulate. To paraphrase from one of the interviewees: if one were a sceptic, one might say it is in the interest of the validation contractors to find issues, which will add further validation needs. The problem within the MD domain is that validation needs to meet the requirements of the regulations, which, as the FDA themselves acknowledge, is difficult to define precisely [6].

### 3.2.2 Time to Market

Getting a new product through development as quickly as possible in order to move into the clinical trials and then commercialisation phases is becoming ever more critical to MD companies. Once a launch date has been set, there is very little appetite for delays. As one interviewee put it *"time is generally the biggest priority. Time out weighs cost a lot of the time here"*. This has a consequence for the software development process in that there is little tolerance for changes to process that might introduce risk to compliance and therefore affect time to market. Following one particular process improvement initiative to automate a production line which led to a delay in time-to-market, one interviewee reported *"... the tolerance has gone to zero now for equipment upgrade or equipment strategy causing a delay in product time to market"*.

The concentration on time to market has also created organisational inertia to change within the new product introduction process. This may be typical within a MD environment, [27] states: "In such a world, any significant change to culture or process can be difficult. Sometimes it is difficult because of the inertia inherent in a large organization". During one interview it was suggested that they could possible improve their manufacturing process if they spent more time researching newer technologies and techniques *"Time to market is the big deal here ... It's very rare that we get to develop prototype equipment, [and] test it"*. This includes the SDLC, because they interact with the development group and would see an iterative approach being more effective for them as opposed to the 'waterfall' type approach which their process requires. What has therefore happened is they have found ways to minimise the process overhead, bending the rules of the process along the way.

# 4 Discussion

Our investigations have brought to light some interesting effects that certain factors, such as regulations, have on the software development process within a MD plant. A possible outcome of having to adhere to regulatory requirements can be the corporate enforcement of a process. Speaking from the perspective of a MD firm, [27] states: "In our safety-critical world, we strongly believe that a robust process is an important element to insuring high quality. The side effect of that strong belief can be an over-reliance on prescriptive, mandated process rules that take the approach of imposing discipline upon a team". Although it is good to have a well defined process, it seems that it is easy to over shoot the mark. A process which is not flexible enough to allow for all aspects of the business, for example the R&D section which requires an iterative type of process, will result in frustrated employees and possibly short cutting of processes.

## 4.1 Leaning the Software Development Process

It is important to have a work ethic of continuous improvement within an organisation. Many process improvement models such as CMMI (level 5) [28], ISO/IEC 15504-5 [29] refer to the highest level of process maturity having continuous improvement as a core element. This is similar to the lean principle of striving for *perfection* [13]. This ethos is evident within MedTech, which has been proactive in tackling the issues by various process improvement initiatives. For example, conscious of the issue of task switching when quotes are needed for potential projects, a process MedTech introduced that helps reduce this, is the availability of an internal web based tool. The tool poses a number of key questions and then gives a high level estimate of the project cost, "... *the number could be out by 20 or 30 % but it's just to give people an order of magnitude*". This allows managers to gauge whether or not to pursue the project further, thus reducing the number of non-value-add project estimates the developers get involved in.

The same approach can also work with the software developers. One of the theses presented by [30] based on their experience of software process improvement initiatives is that "Developers are motivated for change; if possible, start bottom-up with concrete initiatives." There are various tasks the MedTech developers have identified that could lead to process efficiencies, such as better code re-use, code reviews, use of testing tools, and skills development. Many such tasks are independent of the regulatory environment they operate in, however what is needed is some form of organisational support to find time for employees to execute on these initiatives. Within the lean manufacturing world, this is achieved via policy deployment [13] pg. 95: "The idea is for top management to agree on a few simple goals for transitioning [and] to designate the people and resources for getting the projects done". Some software development changes however, could be problematic in a MD company, for example the practice of refactoring source code [31] may cause unexpected results to previously completed (and possibly certified) code [32], [33]. A certain amount of caution is required.

Hiring contract staff is not an uncommon approach to manage surges in capacity, and with Global Software Development [34] becoming common place in many large companies, one could imagine that eventually it will become more common within the MD domain also. The case study has shown that cost considerations are less important than in other industries, especially when it comes to new product R&D. However the long project timelines typical of a new MD generate very large costs, something the MD industry is not immune to. In tackling the huge expense of hiring contract validation engineers, two lean process changes helped to reduce this cost exposure in MedTech. Firstly in relation to the sourcing of validation engineers: *"... we've reduced the numbers from a contractor base and they're not as pernickety now"*. Secondly, taking what can be seen as a lean approach to contract management [17], they changed the contract terms from time and materials to fixed price, and any additional work would have to be renegotiated with the requesting manager who would be far more reluctant to incur additional project costs: *"Now they're very effective at managing scope"*.

A big question researchers should be asking is how to apply software process improvement within this type of domain while keeping risk to compliance as close as possible to zero. Some work has been done within embedded-software domains on how to choose the most appropriate methodology [35] and how agile practices should be considered [36]. However, the only model we found relating to mission or life-critical agile adoption [37] describes a stepped approach to deciding which agile practices are suitable depending on the system's characteristics and qualities. While their framework is aimed at any mission or life-critical system, it does not go into much detail about the various regulatory requirements.

Within MedTech there is a great work ethos of process improvement, and within the software development group this is no different. There is acknowledgement that they could be doing things better *"Everyone using this [the process] is quite frustrated by how much time we spend having to document and test stuff .... And it costs a lot of money"*. They are actively engaged in analysing their current process in order to *"make [it] more effective, efficient and more lean than what we do now"*. This is easier said than done, and especially within the SDLC.

Lean, having its origins in a manufacturing setting, has had limited success in penetrating areas such as software development. However, using the concept of lean software development we can see that MedTech has instigated some practices that have made some initial process improvements. Information hiding through modularisation and componentisation which remove complexity, parameterisation making modules less implementation specific, and code re-use are suggested lean practices [17]. MedTech seem to be going in the right direction having begun to build a library of software objects and re-usable components. As one interviewee attested *"From the testing side I find it personally a lot better because I know that that piece of code, I don't have to check it, it's been done already and working reliably"*.

They have also reduced waste in terms of communication between the software group and the more equipment focused group. Following a recent re-organisation it was reported that *"those two groups have merged under the one manager an there's been a lot more interaction*". Also, the developers themselves have embarked on their own initiative to cross-train each other so as to increase the level of expertise within the group *"Now we interchange rolls so those guys sometimes they do machines and sometimes ... or whoever will do databases, screens, something like that"*.

Further possibilities will be worth exploring for MedTech, such as establishing a TDD approach [20]. With their hardware dependencies, TDD, combined with the use of testing techniques which decouple the hardware such as bracketing-out, mocks, and stubs [38], and hardware simulators [39] could offer process improvements well worth pursuing.

## 5 Conclusion and Future Work

Having examined the SDLC within a MD manufacturing plant, we identified a number of key influencing factors (Fig. 2) and expanded on two in this paper, namely regulations and business focus. While many MD companies are pre-occupied with achieving regulatory compliance, we have seen in this example how it can lead to a feeling of over doing it by applying a heavy process to all aspects of the software life-cycle. We also identified some of the key business drivers (Fig. 3) and again expanded on just two of these: cost and time-to-market. The MD industry appears to becoming much more competitive and cost focused, and therefore companies are looking at ways to improve their processes but without affecting regulatory compliance. Since there does not seem to be a mechanism for quantifying just how much process is enough, it would be beneficial for a focused assessment of existing processes which could indicate where too much rigour is being applied and therefore the possibility to reduce the amount of work required.

Lean software development, although still not very well defined, offers the potential for transferring the principle of lean manufacturing into software development and thereby achieving some of the benefits seen by lean initiatives in other parts of the organisation. What would be useful is a reference model of lean software practices which can be employed while not affecting regulatory compliance.

The MD domain is seen as an area with huge growth potential particularly relevant within Ireland [40]. As an Irish based software engineering research centre, we therefore find it very relevant to conduct research into this domain. Indeed the outlook for this domain, in a global sense, is one of great advancements in technology leading to smaller, more complex devices merged with physiological, biological, engineered and physical systems. Importantly, it is anticipated that the current software development methodologies for such nanoscale systems will have to fundamentally change [41]. The challenge for the research community, therefore, is to develop architectures and methodologies appropriate to supporting these advancements while keeping an open mind as to how the regulatory bodies are likely to respond. Consequently, any process improvement, assessment, or reference model should be future proofed to allow for and support such innovations.

**Acknowledgements.** This research is supported by the Science Foundation Ireland (SFI) Stokes Lectureship Programme, grant number 07/SK/I1299, the SFI Principal Investigator Programme, grant number 08/IN.1/I2030 (the funding of this project was awarded by Science Foundation Ireland under a co-funding initiative by the Irish Government and European Regional Development Fund), and supported in part by Lero - the Irish Software Engineering Research Centre (http://www.lero.ie) grant 03/CE2/I303_1.

## References

1. EU, Council Directive 93/42/EEC of the European Parliament and of the Council, Concerning Medical Devices, E. Council, Editor, Official Journal of the European Union (1993)
2. EU, Directive 2007/47/EC of the European Parliament and of the Council. Official Journal of the European Union (2007)
3. FDA. U.S. Food and Drugs Administration, http://www.fda.gov (accessed March 10, 2011)
4. RTCA, DO-178B, Software Considerations in Airborne Systems and Equipment Certification. RTCA (Radio Technical Commission for Aeronautics) (January 1992)
5. FDA, Code of Federal Regulations 21 CFR Part 820, U.F.a.D. Administration, Editor (April 2009)
6. CDRH, General Principles of Software Validation; Final Guidance for Industry and FDA Staff, FDA (2002)
7. IEEE, IEEE Reliability Society - Annual Technical Report (2008)
8. ANSI/AAMI/IEC, 62304:2006. Medical Device Software-Software life cycle processes, Association for the Advancement of Medical Instrumentation, p. 67 (2006)
9. Spence, J.W.:There has to be a better way! In: AGILE Conference 2005. Inst. of Elec. and Elec. Eng. Computer Society, Denver, CO, United states, July 24–25 (2005)
10. Bosch, T.: Medical Device Software Development—Going Agile, http://www.mddionline.com/article/medical-device-software-developmentmdashgoing-agile (accessed March 10, 2011)
11. Womack, J.P., Jones, D.T., Roos, D.: The Machine That Changed The World: How lean production revolutionized the global car wars. Simon & Schuster (2007)
12. Liker, J.: The Toyota Way 2003. McGraw-Hill, New York (2003)
13. Womack, J.P., Jones, D.T.: Lean Thinking: Banish Waste and Create Wealth in Your Corporation. Simon & Schuster (1996)
14. Cockburn, A.: Agile Software Development. In: Highsmith, C.a. (ed.) The Agile Software Development Series (2002)
15. Highsmith, J.: Agile Software Development Ecosystems. Addison-Wesley, Reading (2002)
16. Beck, K.: Test Driven Development: By Example. Addison-Wesley Professional, Reading (2002)
17. Poppendieck, M., Poppendieck, T.: Lean Software Development: An Agile Toolkit. Addison-Wesley Professional, Reading (2003)
18. Hibbs, C., Jewett, S.C., Sullivan, M.: The Art of Lean Software Development. O'Reilly Media, Sebastopol (2009)

19. AgileAlliance. Manifesto for Agile Software Development, Available from: 10, 20, http://www.agilemanifesto.org/ (accessed March 2011)
20. Rottier, P.A., Rodrigues, V.: Agile Development in a Medical Device Company. In: Agile, 2008. AGILE (2008)
21. Lin, W., Fan, X.: Software development practice for FDA-compliant medical devices. In: 2009 International Joint Conference on Computational Sciences and Optimization, CSO 2009. IEEE Computer Society, Sanya (2009)
22. Cawley, O., Wang, X., Richardson, I.: Lean/Agile Software Development Methodologies in Regulated Environments – State of the Art. In: Proceedings of Lean Enterprise Software and Systems, Helsinki, Finland, October 2010, pp. 31–36. Springer, Heidelberg (2010)
23. Yin, R.K.: Case Study Research: Design and Methods Paperback edn. In: Applied Social Research Methods, Sage Publications, Inc., Thousand Oaks (2009)
24. Miles, M.B., Huberman, M.A.: Qualitative Data Analysis: An Expanded Sourcebook, 2nd edn. Sage Publications, Inc., Thousand Oaks (1994)
25. Strauss, A., Corbin, J.: Basics of qualitative research: grounded theory procedures and techniques. Sage Publications, Thousand Oaks (1990)
26. State Economic Impact of the Medical Technology Industry. The Lewin Group, Inc, http://www.lifechanginginnovation.org/ (accessed March 10 (2010)
27. Weyrauch, K.: What are we arguing about? A framework for defining agile in our organization. In: Agile Conference (2006)
28. CMMI, Capability Maturity Model Integration V1.3, http://www.sei.cmu.edu/cmmi (Accessed March 10, 2011)
29. ISO/IEC 15504-5: 2006. Information Technology – Process Assessment – Part 5: An exemplar Process assessment model (2006)
30. Conradi, H., Fuggetta, A.: Improving software process improvement. In: Software, vol. 19(4), pp. 92–99. IEEE, Los Alamitos (2002)
31. Fowler, M.: Refactoring-Improving The Design Of Existing Code.Hardback edn. In: Booch, Jacobson, Rumbaugh (eds.) Object Technology Series, Addison-Wesley, Reading (2000)
32. Chisholm, R.A.: Agile Software Development Methods and DO-178B Certification. In: Division of Graduate Studies and Research, Royal Military College of Canada (2007)
33. Ronkainen, J., Abrahamsson, P.: Software development under stringent hardware constraints: do agile methods have a chance? In: Marchesi, M., Succi, G. (eds.) XP 2003. LNCS, vol. 2675. Springer, Heidelberg (2003)
34. Herbsleb, J.D., Moitra, D.: Global Software Development. IEEE Software 18(2), 16–20 (2001)
35. Kettunen, P., Laanti, M.: How to steer an embedded software project: tactics for selecting the software process model. In: Information and Software Technology, vol. 47, pp. 587–608. IEEE, Los Alamitos (2005)
36. Srinivasan, J., Dobrin, R., Lundqvist, K.: State of the Art in Using Agile Methods for Embedded Systems Development. In: 33rd Annual IEEE International Computer Software and Applications Conference, COMPSAC 2009 (2009)
37. Sidky, A., Arthur, J.: Determining the Applicability of Agile Practices to Mission and Life-Critical Systems. In: Proceedings of the 31st IEEE Software Engineering Workshop. IEEE Computer Society Press, Los Alamitos (2007)
38. Van Schooenderwoert, N., Morsicato, R.: Taming the embedded tiger - Agile test techniques for embedded software. In: Proceedings of the Agile Development Conference, DCA, Lake City, UT, June 22 - June 26, S 2004. IEEE Computer Society Press, Salt Lake City (2004)

39. Mueller, G., Borzuchowski, J.: Extreme embedded a report from the front line. In: OOPSLA 2002, Practitioners Reports. ACM, Seattle (2002)
40. Expert-Group-on-Future-Skills-Needs, Future Skills Needs of the Irish Medical Devices Sector (2008), http://www.skillsireland.ie/publication (accessed March 10, 2011)
41. High Confidence Software and Systems Coordinating Group, High-Confidence Medical Devices: Cyber-Physical Systems for 21st Century Health Care - A Research and Development Needs Report. The Federal Networking and Information Technology Research and Development (NITRD) Program (2009)

# Standalone Software as an Active Medical Device

Martin McHugh, Fergal McCaffery, and Valentine Casey

Regulated Software Research Group, Dundalk Institute of Technology & Lero,
Dundalk Co. Louth, Ireland
`{Martin.McHugh,Fergal.McCaffery,Val.Casey}@dkit.ie`

**Abstract.** With the release of the latest European Medical Device Directive (MDD) standalone software can now be classified as an active medical device. Consequently the methods used to ensure device safety and reliability needs to be reviewed. IEC 62304 is the current software development lifecycle framework followed by medical device software developers but important processes are beyond the scope of IEC 62304. These processes are covered by additional standards. However since the MDD became mandatory these additional standards are not comprehensive enough to ensure the reliability of an active medical device consisting of only software. By employing software process improvement techniques this software can be developed and validated to ensure it performs the required task in a safe and reliable way.

**Keywords:** Medical Device Standards, IEC 62304, MDD (2007/47/EC), Software Process Improvement.

## 1 Introduction

The use of technology within healthcare is on the rise particularly [1] with the advent of healthcare software applications for use with smartphones such as "Medscape" and "Radiology 2.0" for the Apple iPhone. The iTunes App Store has a section containing over two hundred and thirty applications for use within healthcare [2]. Failures in software used within healthcare can have costly and deadly consequences. This occurred in 2000 when twenty one Panamanian teletherapy patients received lethal doses of radiation therapy due to faulty software [3]. The Food and Drugs Administration (FDA) record all product recalls and in the period from 1st November 2009 to 1st November 2010 seventy eight devices were recalled due to software related problems [4].

An increasing number of tasks within the medical profession are being transferred to automated software driven devices. This can be seen in USB blood glucose meters, where traditionally the measurement was taken by a clinician and the results manually recorded either via pen and paper or entered into an electronic health record (EHR), whereas now the sample is placed on the USB glucose meter and the device automatically records the results and once connected to a computer automatically updates the patient's EHR [5].

Medical devices intended for use within the European Union must have a CE conformance mark [6]. To achieve this conformance mark audits are performed on these devices to ensure their safety and reliability by notified bodies within each country. Within the Republic of Ireland the National Standards Authority of Ireland (NSAI) is responsible for ensuring conformity before awarding a CE mark. These devices typically needed to satisfy standards which include: EN ISO 13485:2003 (medical device quality management standard) [7], EN ISO 14971:2009(medical device risk management standard) [8] and the medical device product level standard IEC 60601-1 [9, 10]. The original directive (MDD 93/42/EEC) historically defined a medical devices as being hardware with or without a software element [11]. However since the enforcement of European Medical Device Directive (MDD 2007/47/EC) in March 2010 [12], standalone software can now be classified as an active medical device [13] and consequently the standards used to ensure conformance to the CE mark need to be reviewed and if necessary amended. As a result of this update to the MDD, medical device software is required to be developed through adopting best practice software development practices. This essentially means adhering to the medical device software lifecycle process standard IEC 62304 [14] and the set of aligned medical device standards and technical reports e.g. IEC 62366 [15] and IEC TR 80002-1 [16].

Software process improvement (SPI) is not a new concept but is becoming increasingly important in the area of medical device software development. The ISO 15504-5 [17] standard also known as SPICE (Software Process Improvement Capability dEtermination) is recognised as a source of best practices for software development projects. SPICE was not developed for any specific sector of the software industry so it is general in its approach.

IEC62304:2006 is a software development lifecycle for use within the medical device software development domain and is derived from the generic software lifecycle process ISO 12207 [18] [19]. IEC 62304 is a harmonised standard since November 2008 with the following European Council medical device standards: MDD (1993/42/EEC); AIMD (1990/385/EEC); and IVDD (1998/79/EC) [20]. Within this paper we examine IEC 62304:2006 to determine if all the requirements of the MDD (2007/47/EC) are satisfied and if not, what framework must be applied in order for software development projects to meet the CE conformance requirements.

The remainder of this paper is structured as follows:

In section 2, the revision of the MDD (2007/47/EC) is examined to see what affect this amendment to the MDD (1993/42/EEC) has on the classification of a medical device and how this classification effects the development of medical device software. Section 3 discusses the importance of SPI. Additionally the history of IEC62304:2006 is analysed and how it is has evolved from the ISO 12207 and SW68 standards [19] along with what processes are included in IEC 62304:2006. Section 4 examines the existing medical device software development standard IEC 62304:2006 to determine if it is comprehensive enough to satisfy the MDD amendment in relation to standalone software now being defined as an active medical device. Section 5 discusses practices beyond the scope of the IEC62304:2006 standard that are required to satisfy the definition of software in the MDD (2007/47/EC). Additionally, we determine whether missing practices may be resolved by amending or extended, the existing processes or if there is a need to develop a new standalone medical device software lifecycle standard. Finally section 6 provides the conclusions from this research and plans to progress this work further.

## 2 Medical Device Directive (2007/47/EC)

All medical devices intended for use within the European Union must conform to the current MDD. The MDD (2007/47/EC) is the current directive and was released on October 11$^{th}$ 2007. However it only became mandatory for CE compliance on March 21$^{st}$ 2010 [21]. The MDD (2007/47/EC) is harmonised with a number of standards relating to the production of medical devices e.g. EN ISO 14971:2009 and IEC EN 62304:2006. MDD (2007/47/EC) Article I Section 2 defines a medical device as [13]:

*"any instrument, apparatus, appliance, **software**, material or other article, whether used alone or in combination, including the software intended by its manufacturer to be used specifically for diagnostic and/or therapeutic purposes and necessary for its proper application"*

As highlighted in the above definition as to what constitutes a medical device, standalone software can be considered a medical device. Whilst MDD (1993/42/EEC) did allow for software to be seen as a medical device [11] it did not extend to standalone software being recognised as an active medical device. MDD (2007/47/EC) Annex IX Section 1.4 defines an active medical device as [13]:

*"any medical device operation of which depends on a source of electrical energy or any source of power other than that directly generated by the human body or gravity and which acts by converting this energy. Medical devices intended to transmit energy, substances or other elements between an active medical device and the patient, without any significant change, are not considered to be active medical devices. Stand-alone software is considered to be an active medical device"*

Methods used to ensure device conformity to the MDD (1993/42/EEC) have not been modified with the release of MDD (2007/47/EC) even though the definition of a medical device has changed with particular reference to standalone software being capable of being an active medical device An example of software as an active medical device is software used to plan cancer treatment doses and to control the setting of oncology treatment devices

### 2.1 Classification Rules

Annex IX section III (Classification) within the MDD (2007/47/EC) categorizes medical devices into one of four categories [13]:

➢ Class I devices are non-invasive devices unless they are used for the purpose of channelling blood or tissue or unless they are intended for use on wounds which have the dermis breached and can only heal by secondary intent, e.g. wheelchairs, bandages, incontinence pads. Also invasive devices that are not connected to an active medical device are classified as Class I e.g. tongue depressor.
➢ Class IIa devices are surgically invasive devices for transient use unless they control, diagnose, monitor or correct a defect of the heart of central circulatory system e.g. transfusion equipment, storage and transport of donor organs.

> Class IIb devices are surgically invasive devices which are implantable or intended for long term use unless they come into contact with the heart e.g. haemodialysis, dressings for chronic extensive ulcerated wounds.
> Class III devices are used for supporting or sustaining life and devices which potentially pose an unreasonable risk of illness or injury e.g. pacemakers and heart valves.

This classification is based on the risk to the patient's safety, ranging from low risk to high risk. The higher the risk to the patient's safety, the greater the level of assessment required to achieve the CE conformance mark. If software is part of a medical device it assumes the classification of the overall device. If standalone software is an active medical device then the device is classified based on the risk the device places on the patient or a third party according to MDD (2007/47/EC) Annex IX Section III [13].

The MDD (2007/47/EEC) has wider reaching consequences, devices that historically were not classified as medical devices and not subject to conformance standards are now being classified as medical devices. This occurs when the device is connected to an active medical device. An example of this is in the visual display unit (VDUs) that display results from a medical device are now classified as being a medical device [22].

## 3 Software Process Improvement & IEC 62304:2006

### 3.1 Software Process Improvement

SPI is an important element of any software development project. SPI is a continuous cyclic path of improvement by performing assessments, implementing recommendations of those assessments and beginning the cycle again [23]. All software development projects are a series of processes. The processes need to be understood and improved where possible. There are four primary objectives of SPI [24]:

> To attain an understanding of current software development practices;
> Select areas to focus on to achieve the greatest long term benefits;
> Add value to the organisation developing the software rather than solely to a specific development project;
> To grow by combining effective processes with skilled and motivated people.

Private industry has greatly benefited from SPI. Hughes Aircraft invested $400,000 to develop software process improvement within their company. This investment resulted in an annual saving of $2,000,000 to the company. Similarly safety is improved by employing SPI practices. An improvement process was undertaken by the group that develops on-board software for the Space Shuttle at IBM Huston. Early detection rate of errors rose from 48% to 80% as a result of using SPI [25].

A recent survey carried out by Embedded Market Surveys analysed software development projects in both embedded industry and the medical device industry [26]. The survey found that projects within embedded industry on average over run by 47%

whilst software development projects within medical device production over run by 54.4%. The survey also found that 9.7% of medical device projects were cancelled with the reasons being cited:

- Incomplete or vague requirements;
- Insufficient time;
- Insufficient resources;
- Design Complexity.

These issues can be overcome by employing an effective software development process which can then reduce project over runs, cost can be reduced and the number of projects cancelled can be decreased. SPI frameworks such as ISO 15504-5, and CMMI® [27] and international standards such as ISO\IEC 12207 or IEC 62304 provide guidance on how to help address all of the above areas either directly or by using normative reference to additional standards.

## 3.2 IEC 62304:2006 Medical Device Software – Software Lifecycle

IEC 62304:2006 is derived from ISO/IEC 12207, by the Association of Advancement of Medical Instrumentation (AAMI) and from the American National Standards ANSI/AMMI SW68:2001 [19].

Whilst ISO/IEC 12207 is not domain specific it is seen as being very comprehensive in its approach and this is reflected from the standards utilising the core principle of ISO/IEC 12207 e.g. ISO IEC 15504-5:2006 and ISO/IEC 90003. IEC 62304 was developed between 2002 and 2006 by the joint working group ISO/TC 210 and IEC Sub-Committee 62A [28]. IEC 62304 was created as a software development standard for lifecycle processes for the purpose of safe design and maintenance of medical device software.

Software developed using the IEC 62304:2006 standard is founded upon the assumption that the software is developed in accordance with a quality management standard (ISO 13485:2006), a risk management standard (ISO/IEC 14971) and a product level standard (EN 60601-1) [9].

IEC 62304:2006 section 3.24 defines a Software Lifecycle Model as [14]:

*A conceptual structure spanning the life of the software from definition of its requirements to its release for manufacturing which:*

➢ *Identifies the process, activities and tasks involved in development of a software product*
➢ *describes the sequence of and dependency between activities and tasks*
➢ *identifies the milestones at which the completeness of specified deliverables is verified*

This definition is based on the definition within ISO/IEC 12207:1995, definition 3.11 [18].

This standard provides a framework of process with activities and tasks. A process is divided into activities and the activities are further divided into tasks. The processes within IEC 62304:2006 for the development of software for medical devices are [14]:

- Quality Management System
- Software Safety Classification
- Software Development Process which includes;
  - Software Development Planning
  - Software Requirements Analysis
  - Software Architectural Design
  - Software Detailed Design
  - Software unit implementation and verification
  - Software Integration and testing
  - Software system testing
  - Software Release
- Software Maintenance Process
- Risk Management Process
- Software Configuration Management Process
- Software Problem Resolution Process

These processes are represented graphically in Figure 1 and it can be seen how these processes fit into IEC 62304:2006.

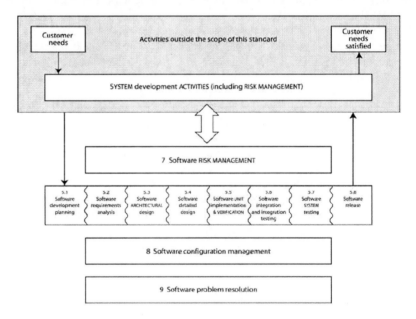

**Fig. 1.** Overview of Software Development Processes and Activities [14]

Within IEC 62304:2006 Section 4.3 [14], software is classified according to the severity of potential harm, into one of three categories similar to that of the medical device classification:

- Class A: No injury or damage to health is possible
- Class B: Non-serious injury is possible
- Class C: Death or Serious Injury is possible

These classifications are subject to the medical device risk management standard ISO 14971:2007. As the risk management process is covered by ISO 14971:2007, IEC 62304:2006 makes normative reference to it. While making some additions which are required for the identification of software factors related to hazards.

Safety critical standalone software systems can be divided into items running a different software element each with its own safety classification. These items can be further sub divided into additional software elements. The overall software system assumes the highest classification contained within all of the software elements. For example if a software system contains five software elements, four of which may be classified as Class A, but one may be classified as Class C and therefore the overall device receives a classification of Class C [29]. This can be seen in figure 2.

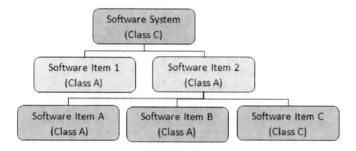

**Fig. 2.** Classification of software items within complete software system

## 4 Problems with Existing Conformance Standards

As discussed in section two, the definition of what constitutes a medical device has changed under MDD (2007/47/EC), however the methods used to test these devices has not changed in response to the latest directive amendment.

The current standard IEC 62304:2006 operates in conjunction with additional standards such as EN ISO 14971:2007 and EN ISO 13485:2003 to ensure overall conformance of the device. A number of processes have explicitly been defined as being beyond the scope of IEC 62304:2006

Figure 1 shows that there are processes beyond the scope of IEC 62304:2006 such as:

- Customer needs (Requirements elicitation)
- System Development Activities including risk management
- Customer needs satisfied (Validation and Final Release)

IEC 62304 was developed prior to MDD (2007/47/EC) and relies on additional standards to ensure the safety and reliability of the overall system. For this standard to be seen as a comprehensive method of ensuring the safety and reliability of standalone software as an active medical device these processes need to be brought into the scope.

Processes beyond the scope of IEC 62304 prior to MDD (2007/47/EC) were performed within other standards, but these standards do not address the specific aspects relating to developing standalone software as an active medical device. Essentially IEC 62304 was designed for medical device software that would always be part of a traditional medical device system (i.e. hardware and software) and therefore the focus was only on the software component and the system level processes and practices are not addressed. However, as the MDD defines standalone software as an active medical device the medical device system consists solely of software. Therefore to meet the requirements of the revised MDD system level processes and practices need to be handled and this is not currently possible using IEC 62304 in isolation.

## 5 Resolving the Problems

Research performed by the authors as (part of the Regulated Software Research Group (RSRG)) in Dundalk Institute of Technology (DkIT) has identified the following list of processes that need to be completed to ensure the development of safe and reliable software as an active medical device:

- Software System Project Management Planning;
- Software System Requirements Elicitation;
- Software System Requirements Analysis;
- Software System Architectural Design;
- Software System Design;
- Software System Construction;
- Software System Integration;
- Software System Testing;
- Software System Release;
- Software System Installation;
- Software System Maintenance.

These processes correspond to the key stages required for medical device software development. Each process has been designed to address the requirements of developing software as a standalone system. The following supporting processes have also been identified:

- Validation;
- Configuration Management;
- Change Request Management;
- Problem Resolution Management.

IEC 62304:2006 incorporates the majority of these processes. Processes not covered by IEC 62304:2006 include: Software System Requirements Elicitation; System Installation; Validation; and System Final Release (Software System Release Process only deals with software Release and not final release). The processes not covered by IEC 62304:2006 are however covered by ISO/IEC 15504-5 but as discussed ISO/IEC 15504-5 is not domain specific. Consequently, the processes not included within IEC 62304:2006 need to be reviewed and tailored to suit the needs of software being developed as an active medical device.

The processes outside of the scope of IEC 62304:2006 prior to the release of MDD (2007/47/EC) were covered by standards such as EN ISO 13485:2003, ISO 14971:2007 and EN 60601-1:1990 etc. Currently within IEC 62304:2006 there is no method of validating a complete system that only incorporates software. This results in the standard not being capable of ensuring the safety and reliability of software that is an active medical device.

Device manufacturers within other industries have had similar problems and as a result they developed standards which meet their needs and strove towards best practice rather than simply reaching conformance requirements. These include Automotive Spice and Spice for Space, both of which allow for process assessment and provide an improved software development path [23]. Similarly within the medical device industry Medi SPICE is being developed to provide a basis for medical device software process assessment and improvement. Medi SPICE aims to provide developers of medical device software with a complete lifecycle to develop medical device software. This includes conformance with regulatory requirements and the processes beyond the scope of IEC 62304. It is therefore envisaged that it can be used to help achieve CE compliance of standalone software as an active medical device. The work we present in this paper is being factored into Medi SPICE so that both embedded medical device software development and standalone medical device software will be covered [30].

## 6 Conclusions

For a medical device manufacturer to be successful, it must produce a safe, reliable device that conforms to the regulatory standards of the market into which they are selling their devices i.e. MDD (2007/47/EC) within the European Union. Upon achieving conformance medical device manufactures can market and sell their products within a particular market. Conformance to these standards ensures that only safe and reliable products are used. The MDD (2007/47/EC) has incorporated a number of changes in comparison to MDD (1993/42/EEC). In terms of this paper the most significant change within MDD (2007/47/EC) is the ability of software to be now seen as an active medical device. But since this amendment the question of whether or not the existing standards used to regulate this software are sufficient must be answered.

IEC 62304:2006 is the current standard used by software developers developing software for medical devices. This standard is part of the harmonised standards within MDD (2007/47/EC). IEC 62304:2006 provides a framework of lifecycle processes. However there are important system processes that are beyond the scope of IEC 62304:2006

The processes outside of the scope of IEC 62304:2006 are primarily system processes which reference ISO/IEC 12207. As ISO/IEC 12207 is a generic software lifecycle rather than being domain specific it can be considered broad and does not fully cater for the needs of medical device software development. These processes need to be tailored to suit the needs of medical device software. This will be achieved through the release of Medi SPICE.

**Acknowledgements.** This research is supported by the Science Foundation Ireland (SFI) Stokes Lectureship Programme, grant number 07/SK/I1299, the SFI Principal Investigator Programme, grant number 08/IN.1/I2030 (the funding of this project was awarded by Science Foundation Ireland under a co-funding initiative by the Irish Government and European Regional Development Fund), and supported in part by Lero - the Irish Software Engineering Research Centre (http://www.lero.ie) grant 03/CE2/I303_1.

## References

1. Blacksmith Enterprise, http://www.blackenterprise.com/2010/09/30/healthcare-it-on-the-rise/
2. PCMag.com, http://www.pcmag.com/article2/0,2817,2343550,00.asp
3. Borrás, C.: Overexposure of Radiation Theraphy in Panama. Articles and special reports, 173–187 (2006)
4. U.S. Department of Health & Human Service, http://www.accessdata.fda.gov/scripts/cdrh/cfdocs/cfres/res.cfm
5. Diabetes. Bayer Contour USB Glucose Meter (2010), http://www.diabeticlive.com/glucose-meters/bayer-contour-usb-glucose-meter/
6. British Standards Online (America), http://www.bsiamerica.com/en-us/Sectors-and-Services/Industry-sectors/Healthcare-and-medical-devices/CE-marking-for-medical-devices/
7. EN ISO 13485:2003. Medical Device: Quality Management Systems. Requirements for the Regulatory Process, July 24 (2003)
8. EN ISO 14971:2009. Medical Devices. Application of Risk management to medical devices, July 31 (2009)
9. EN 60601-1. Medical Electrical Equipment. General requirements for basic safety and essential performance. Collateral standard. Usability, May 31 (2010)
10. Emergo Group, http://www.bsiamerica.com/en-us/Sectors-and-Services/Industry-sectors/Healthcare-and-medical-devices/CE-marking-for-medical-devices
11. European Council, Council Directive 93/42/EEC Concerning Medical Devices, June 14 (1993)
12. Webb, K.: Changes to the Medical Device Legislation in Europe- The Impact of Directive, 2007/47/EC (2007), http://www.slideshare.net/mediqol/changes-to-the-medical-device-legislation-in-europe-presentation
13. European Council 2007, Council Directive 2007/47/EC (September 2007)

14. ANSI/AAMI/IEC 62304. Medical device Software - Software life cycle processes. Association for the Advancement of Medical Instrumentation, July 19 (2006)
15. IEC 62366. Medical Devices - Application of usability engineering to medical devices, November 13 (2007)
16. AAMI/IEC TIR 80002-1:2009. Medical Device Software 1: Guidance on the application of ISO 14971 to Medical Device Software, May 31 (2010)
17. ISO/IEC 15504-5:2006. Information Technology. Process Assessment. An exemplar process assessment model, April 28 (2006)
18. ISO/IEC 12207:1995. Information Technology. Software Lifecycle Processes, February 28 (1995)
19. Qmed, `http://www.qmed.com/consultants/19204/global-perspectives-north-america-iec-62304-questions-and-answers?quicktabs_5=1`
20. Eagles, S.: International Standards and EU regulation of medical device software – An update (2009)
21. Emergo Group, `http://www.emergogroup.com/newsletters/directive-2007-47-ec-sep2007`
22. ComputerWorld. FCC mobile network plan could revolutionize health care (2010), `http://www.computerworld.com/s/article/9174429/FCC_mobile_network_plan_could_revolutionize_health_care`
23. McCaffery, F., Coleman, G.: The need for a software process improvement model for the Medical Device Industry. In: International Review on Computers and Software, vol. 2, pp. 10–15 (2007)
24. Wiegers, K.: Software Process Improvement: Ten Traps to avoid, `http://www.processimpact.com`
25. Humphrey, W.S.: Introduction to Software Process Improvement. Technical Report CMU/SEI-92-TR-007 ESC-TR-92-007
26. Embedded Forecasters - Embedded Market Forecasters Survey (2010)
27. CMMI Product Team, Capability Maturity Model® Integration for Development Version 1.2, Software Engineering Institute (2006)
28. Miura, S.: Industry View Point on software In: 11th Conference of the Global Harmonisation Task Force (2007)
29. Gerber, C.: Introduction to IEC 62304 Software Life cycle for medical devices, September 4 (2008)
30. McCaffery, F., Dorling, A.: Medi SPICE: An Overview. Journal of Software Maintenance and Evolution: Research and Practice 22, 255–267 (2010)

# Assessment of Software Process and Metrics to Support Quantitative Understanding: Experience from an Undefined Task Management Process

Ayca Tarhan[1] and Onur Demirors[2]

[1] Hacettepe University Computer Engineering Department
Beytepe Yerleskesi, 06532, Ankara, Turkey
atarhan@cs.hacettepe.edu.tr
[2] Middle East Technical University Informatics Institute
Inonu Bulvari, 06531, Ankara, Turkey
demirors@metu.edu.tr

**Abstract.** Software engineering management demands the measurement, evaluation and improvement of the software processes and products. However, the utilization of measurement and analysis in software engineering is not very straightforward. It requires knowledge on the concepts of measurement, process management, and statistics as well as on their practical applications. We developed a systematic approach to evaluate the suitability of a software process and its measures for quantitative analysis, and have applied the approach in several industrial contexts. This paper explains the experience of evaluating a task management process and related measures of a government research agency. The agency had not defined the task management and measurement processes, and the performance data were gathered from a change management tool. We spent six person-days performing the assessment and analyzing data from 92 process executions. We observed that as systematic approaches have become available, software organizations are able to readily apply quantitative techniques.

**Keywords:** Software measurement, quantitative analysis, control chart.

## 1 Introduction

We need to understand quantitatively the current status of what we are dealing with before attempting to manage it. We can use measures for understanding a product's quality, for controlling a project's progress, or for improving a process's performance. Without quantitative understanding, neither effective control nor improvement is possible. Research on quantitative management in software engineering includes distinct applications of various techniques [1,2,3,4,5] as well as high maturity applications [6,7,8,9,10] related to process improvement models like CMM/CMMI [11] and ISO/IEC 15504 [12]. Nevertheless, the studies are scarce.

From a general perspective, the utilization of quantitative techniques in software engineering requires dealing with a number of challenges including process, measurement, and statistics. The process challenge is about understanding the components of a software process that produces data for the quantitative management [13]. The measurement challenge is related to the implementing of software measurement activities as well as selecting the suitable measures for the quantitative analysis [14,15]. The statistics challenge is about applying the correct statistical methods to software measurement data [16,17,18]. In addition, the software domain's inherent characteristics such as people-dependency, creativeness, and changeability [19,20,21] adversely affect these challenges.

From the perspective of the process improvement frameworks, the challenges for quantitative management differ slightly depending on whether the improvement is required at an organizational level or on a per process basis. In either case, dedicated effort of several years are required. Achieving quantitative management at the organizational level demands the definition and satisfaction of the requirements of a number of key practices through maturity levels 4 and 5. Achieving quantitative management on a per process basis, on the other hand, requires the definition and application of the practices through the process capability levels. The challenges of achieving high maturity in CMM/CMMI, for example, have been identified and acknowledged by a number of workshops and surveys carried out particularly after 1999. These include the Survey of High Maturity Organizations in 1999 [22], High Maturity Workshops in 1999 [23] and 2001 [24], State of Measurement Practice Survey in 2006 [25], and CMMI High Maturity Measurement and Analysis Workshop in 2008 [26].

The challenges stated previously and the lack of a generic approach to assess the suitability of a software process for quantitative analysis encouraged us to develop an Assessment Approach for Quantitative Process Management ($A^2$QPM) [27]. The $A^2$QPM enables the systematic identification of a process's inner attributes (e.g. inputs, outputs and activities) and outer factors (e.g. people and the environment) as well as a number of usability characteristics for process measures. We have applied the approach in several industrial contexts [28] where we utilized the control charts as the quantitative analysis technique.

This paper explains the experience of using the approach in evaluating the task management practices and related measures of a government research agency for the achievement of a quantitative understanding. The agency had been developing systems and software for ten years, neither had a defined task management process nor a measurement process, but had been storing process data on various engineering tools. We spent six person-days performing the assessment, analyzing task management data from 92 process executions, and deriving the results. We observed that the task management process that did not include the activity of task verification demonstrated controlled variation, whereas the task management process that included the activity of task verification demonstrated uncontrolled variation, with respect to the duration estimation capability measure. These results were in keeping with the findings of the assessment approach. More importantly, we observed that as systematic approaches have become available, software organizations are able to readily apply quantitative techniques.

The remainder of the paper is organized as follows: Section two provides an overview of the assessment method, section three explains details of the assessment for the task management process and its measures and presents the results, and finally, section four provides the conclusions.

## 2 The Assessment Method

The Assessment Approach for Quantitative Process Management ($A^2QPM$) includes an assessment process that guides the evaluation, an assessment model that defines assets to evaluate a process and measures, and an assessment tool that supports this evaluation. Description of the assessment process can be reached from [27] and features of the assessment tool can be found in [29]. In this section, we briefly explain the assessment model as a base to the implementation explained in section three. The assessment model addresses two issues: Systematic clustering of process executions and data, and evaluating measure and data usability for the quantitative analysis.

Software process data often represent multiple sources that need to be treated separately, and discovering multiple sources requires the careful investigation of process executions. We want to cluster process executions as stemming from a single and constant system of chance causes. For systematic clustering of process executions and data, we developed a method for stratification based on a number of process attributes. Stratification is a technique used to analyze or divide a universe of data into homogeneous groups [30]. Our clustering method operates according to the changes in the values of process attributes such as inputs, outputs, activities, roles, and tools and techniques. If the executions of a process show similarity in terms of these attributes, then we assume that process executions form a homogeneous subgroup (or "cluster" as we call it) which consistently performs among its executions. The process cluster, then, is a candidate for the quantitative control.

Evaluating measure and data usability for the quantitative analysis includes the elaboration of basic measurement practices as well as measurement data existence and measure characteristics. Even if there is not an established measurement process, we can evaluate the practices applied during data collection and analysis. The assessment model identifies a number of measure usability attributes such as measure identity, data existence, data verifiability, data dependability, data normalize-ability, and data integrate-ability. The model defines questionnaires based on these attributes, and recommends investigating a measure's usability for the quantitative analysis prior to an implementation. Each usability attribute is rated in the following four values in the ordinal scale: Fully usable (F), largely usable (L), partially usable (P), and not usable (N). Overall usability of a measure is decided based on the ratings of the attributes.

The assessment model includes assets to evaluate the suitability of a process and its measures for the quantitative analysis. A *Process Execution Record* is used to capture instant values of process attributes for a process execution. Actual values of inputs, outputs, activities, roles, and tools and techniques are recorded on this form for each process execution. A *Process Similarity Matrix* is used to verify the values of process attributes against process executions. The values of process attributes are shown in the rows, and process execution numbers are shown in the columns of the matrix. This shows the differences between the process executions in terms of the process attribute values and enables to identify the clusters of process executions. A *Process Execution Questionnaire* is used to capture the extraordinary cases for a process execution in terms of outer factors such as changes in process performers, process environments, and etc. A *Measure Usability Questionnaire* is used to evaluate the usability of a process measure for the quantitative analysis in terms of measure usability attributes.

## 3 The Implementation

The implementation of the $A^2$QPM included the assessment and analysis of the task management process of a project unit. The unit had been developing systems and software for ten years within a government research agency. The unit had been undertaking projects to develop software for military systems and had 18 staff including the project manager. It had ISO 9001 [31] certificate as related to the agency and had been pursuing process improvement studies to achieve CMMI maturity level 3 for 20 months. The project unit had documented neither a task management process nor a measurement process but had been collecting and storing data by the engineering tools that had been supporting the projects' processes.

The task management process was in use for 16 months via a change management tool for a military project of the unit. Although the process was not officially defined, its steps had been encoded into the tool by a number of task states (so we called the process "undefined"). The states of the task management process are shown in Fig. 1. Every task of the project was being entered into the tool by a task assigner (Project Manager or Team Leader) together with the values of the fields for *task name, task implementer, estimated start date,* and *estimated finish date*. The task implementer, who noticed the assignment, then was starting to work on the task by recording the value for the *actual start date* on the tool. When the task finished, the task implementer was entering the value for the *actual finish date* into the tool. Finally, the task was being closed after being verified by the task assigner.

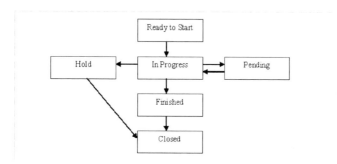

**Fig. 1.** Task states in the task management process

We spent six person-days applying the approach, performing the analyses, and interpreting the results. We worked on existing task management data from 92 process executions which were carried out during 16 months. We identified the attributes of the task management process by inspecting the records entered into the change management tool and consulting the process performers. Since the study was retrospective, we sampled four task records from the set of 92 and completed a Process Execution Record for each. We then gathered a unified set of process attribute values from the sample records and entered these values into the rows of a Process Similarity Matrix as the first column. The rest of the columns, one for each 92 process executions, were utilized to verify the process attributes against the unified set of values in the first column. Whenever a new value for an attribute showed up, it was merged into the values by adding a corresponding row into the matrix.

After verifying process attributes on the Process Similarity Matrix, we analyzed it to identify the process clusters. Every column which was unique in terms of attribute values was identified as a distinct "process cluster". Initially we identified 18 process clusters, out of 92 process executions, each cluster having different combinations of attribute values. Fig. 2 shows the initial clusters.

The number of process clusters was so high that we decided to merge the clusters by considering the purpose of the implementation. The purpose could affect the rules of merging. For example, if the project manager wanted to observe specifically how the process was applied to produce different types of work products, we would use only "outputs" process attribute and merge process clusters having the same types of outputs. Alternatively, if the project manager tried to meet specification limits set by the project's customer on the schedule, we would choose not to merge any of the clusters. In our case, the purpose of the implementation was to understand quantitatively the strong and the weak points in the task management process in order to derive findings for the improvement of the process.

In Fig. 2, we observed that the values of "outputs" attribute were the primary source for the variability between the process executions. Although this finding might be significant to improve the process of developing the work products, it was not that critical to improve the task management process. Therefore, we decided to use "outputs" attribute as being the secondary factor in identifying process clusters rather than being the primary factor. Therefore, we first excluded the values of "outputs" attribute while detecting process clusters, and come up with 4 process clusters labeled from A through D as shown in Fig. 2. We then utilized the values of "outputs" attribute to categorize each of four process clusters into its sub-clusters that were numbered from 1 to 6 with respect to output value type: 1) Document, 2) Software code, 3) Analysis Knowledge, 4) Design, 5) Research Knowledge, and 6) Unclassified output (admin, test, etc.). A graphical representation of the process cluster D in Event Driven Process Chain (EPC) notation is provided in Fig. 3 to visualize the attributes of the task management process.

| Process Attributes | | | PE13 | PE5 | PE31 | PE53 | PE7 | PE28 | PE1 | PE2 | PE16 | PE17 | PE22 | PE12 | PE79 | PE15 | PE20 | PE8 | PE4 | PE60 |
|---|---|---|---|---|---|---|---|---|---|---|---|---|---|---|---|---|---|---|---|---|
| 1 | Inputs | | | | | | | | | | | | | | | | | | | |
| | 1.1 | Task request | o | o | o | o | o | o | o | o | o | o | o | o | o | o | o | o | o | o |
| 2 | Outputs | | | | | | | | | | | | | | | | | | | |
| | 2.1 | Document | | | | | | | o | | | | | | o | | | | | |
| | 2.2 | Software code | o | | o | o | | | | | o | | | | | | o | | | |
| | 2.3 | Analysis knowledge | | o | | | o | | | | | o | | | | | | o | | |
| | 2.4 | Design | | | | | | o | | | | | o | | | | | | o | |
| | 2.5 | Research knowledge | | | | | | | | | | | | o | | | | | | o |
| | 2.6 | Unclassified output | | | | | | o | | | | | | | o | | | | | o |
| 3 | Activities | | | | | | | | | | | | | | | | | | | |
| | 3.1 | Enter task request | o | o | o | o | o | o | o | o | o | o | o | o | o | o | o | o | o | o |
| | 3.2 | Implement task request | | | | | | | o | o | o | o | o | o | | | | | | |
| | 3.3 | Verify task request | | | | | | | | | | | | | o | o | o | o | o | o |
| 4 | Roles | | | | | | | | | | | | | | | | | | | |
| | 4.1 | Task assigner | o | o | o | o | o | o | o | o | o | o | o | o | o | o | o | o | o | o |
| | 4.2 | Task implementer | | | o | o | o | o | o | o | o | o | o | o | o | o | o | o | o | o |
| 5 | Tools and Techniques | | | | | | | | | | | | | | | | | | | |
| | 5.1 | Startteam | o | o | o | o | o | o | o | o | o | o | o | o | o | o | o | o | o | o |
| Process Cluster | | | A | A | B | B | B | B | C | C | C | C | C | C | D | D | D | D | D | D |
| Process Sub-cluster | | | 2 | 3 | 2 | 3 | 4 | 6 | 1 | 2 | 3 | 4 | 5 | 6 | 1 | 2 | 3 | 4 | 5 | 6 |

**Fig. 2.** Process clusters and sub-clusters identified for the task management process by using the Process Similarity Matrix

# Assessment of Software Process and Metrics to Support Quantitative Understanding 113

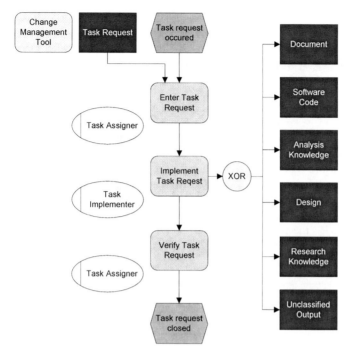

**Fig. 3.** A graphical representation of the task management process cluster D

As the next step, we evaluated the suitability of the task management process measures for the quantitative analysis. We identified *estimated start date*, *estimated finish date*, *actual start date*, and *actual finish date* as base measures of the process. We selected these measures because data were available for them on the change management tool. From the base measures, we first identified *estimated duration* and *actual duration*, as the intermediate derived measures. We then utilized these two measures to derive a final measure, namely the *duration estimation capability*. Fig. 4 shows the measures of the task management process. The arrows in the figure show the relationships between the base measures at the upper side to the derived measures at the lower side.

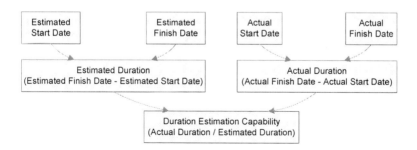

**Fig. 4.** Base and derived measures identified for the task management process

We first completed a Measure Usability Questionnaire (MUQ) for each base measure. By evaluating the answers in the questionnaires, we decided on the usability of the estimated start date, estimated finish date, actual start date, and actual finish date measures in the quantitative analysis. None of the base measures were usable in the quantitative analysis because their data were of interval scale (e.g. we could not take an arithmetic mean on the values such as "21th March, 2008"). In addition, the data dependability attributes for the actual start date and actual finish date measures were evaluated as "partial". An example MUQ for the *actual start date* measure is given in Fig. 5. We also completed a MUQ for each intermediate derived measure and evaluated its usability in the quantitative analysis. The usability of the estimated duration measure was evaluated as "large" whereas the usability of the actual duration measure was evaluated as "partial" for the quantitative analysis. Finally, we completed a MUQ for the duration estimation capability measure and identified that it was "partially usable" in the quantitative analysis. Although this result was below our expectations for the usability of a measure, we decided to continue the quantitative analysis to validate or invalidate the findings of the assessment approach.

| Attributes / Indicators | Answers | Rating | Expected Answers |
|---|---|---|---|
| **Measure Identity** | | N | |
| Q1 Which entity does the measure measure? | Task | | |
| Q2 Which attribute of the entity does the measure measure? | Timing | | |
| Q3 What is the scale of the measurement data? (nominal, ordinal, interval, ratio, absolute) | Interval | | Ratio, Absolute |
| Q4 What is the unit of the measurement data? | Date | | |
| Q5 What is the type of the measurement data? (integer, real, etc.) | Date | | |
| Q6 What is the range of the measurement data? | [3.Jan.2006, 26.Apr.2007] | | |
| **Data Existence** | | F | |
| Q7 Is measurement data existent? | Yes | | |
| Q8 What is the amount of overall observations? | 67 | | |
| Q9 What is the amount of missing data points? | 27 | | Available > 20 |
| Q10 Are data points missing in periods? (If yes, please state observation numbers for missing periods) | No | | |
| Q11 Is measurement data time sequenced? (If no, please state how measurement data is sequenced) | Yes | | |
| **Data Verifiability** | | F | |
| Q12 When is measurement data recorded in the process? (at start, middle, end, later, etc.) | At middle (while accepting task request) | | |
| Q13 Is all measurement data recorded at the same place in the process? (at start, middle, end, later, etc.) | Yes | √ | Yes |
| Q14 Who is responsible for recording measurement data? | Task implementer | | |
| Q15 Is all measurement data recorded by the responsible body? | Yes | √ | Yes |
| Q16 How is measurement data recorded? (on a form, report, tool, etc.) | On a tool (Starteam) | | |
| Q17 Is all measurement data recorded the same way? (on a form, report, tool, etc.) | Yes | √ | Yes |
| Q18 Where is measurement data stored? (in a file, database, etc.) | Starteam database | | |
| Q19 Is all measurement data stored in the same place? (in a file, database, etc.) | Yes | √ | Yes |
| **Data Dependability** | | P | |
| Q20 What is the frequency of generating measurement data? (asynchronously, daily, weekly, monthly, etc.) | Asynchronously (when a new task request is accepted) | | |
| Q21 What is the frequency of recording measurement data? (asynchronously, daily, weekly, monthly, etc.) | Asynchronously (when a new task request is accepted) | | |
| Q22 What is the frequency of storing measurement data? (asynchronously, daily, weekly, monthly, etc.) | Asynchronously (when a new task request is accepted) | | |
| Q23 Are the frequencies for data generation, recording, and storing different? | No | √ | No |
| Q24 Is measurement data recorded precisely? | Yes | √ | Yes |
| Q25 Is measurement data collected for a specific purpose? | Yes (for the purpose of task monitoring) | √ | Yes |
| Q26 Is the purpose of measurement data collection known by process performers? | Yes | √ | Yes |
| Q27 Is measurement data analyzed and reported? | No | | Yes |
| Q28 Is measurement data analysis results communicated to process performers? | No | | Yes |
| Q29 Is measurement data analysis results communicated to management? | No | | Yes |
| Q30 Is measurement data analysis results used as a basis for decision making? | No | | Yes |
| **Data Normalizability** | | | |
| Q31 Can measurement data be normalized by parameters or measures? (If yes, please specify them) | N/A | | |
| **Data Integrability** | | | |
| Q32 Is measurement data integrable at project level? | Yes | | |
| Q33 Is measurement data integrable at organization level? | No | | |

**Fig. 5.** An example Measure Usability Questionnaire for the "actual start date" base measure

We utilized control charts for the quantitative analysis of the task management process with respect to duration estimation capability measure. Control charting is a technique for detecting which type of variation is included in a process. Every process is subject to variation; however, while some processes display *controlled variation*, others display *uncontrolled variation* [30]. Controlled variation means stable and consistent variation over time and consists of common causes. Uncontrolled variation, on the other hand, indicates a pattern that changes over time in an unpredictable manner, and is characterized by special causes. We applied Individuals and Moving Range (XmR) charts to measurement data by using the Minitab Statistical Software [32]. We included the following tests to detect the out-of-control points [30]: (1) 1 point > 3 standard deviations from center line, (2) 9 points in a row on same side of center line, (3) 2 out of 3 points > 2 standard deviations from center line (same side), and (4) 4 out of 5 points > 1 standard deviation from center line (same side).

We reviewed process data and used the results from process similarity assessment and measure usability evaluation to finalize process clusters and measures prior to control charting. Actual start date and actual finish date fields were empty for process cluster A and actual finish date field was empty for process cluster B; therefore, we excluded process clusters A and B from our study. We first charted combined data of process clusters C and D to observe the nature of variation in the task management process with respect to the duration estimation capability measure. The corresponding control charts are shown in Fig. 6(a). There were four out-of-control points (OCPs) in the individuals chart and five OCPs in the moving range chart (marked by squares in the figure). The small number next to an OCP indicates the number of the test violated by the point (or the process execution). An investigation for the reasons of the OCPs appearing in the individuals chart showed that in the executions that violated the test number 1, the closure of the tasks had been forgotten and carried out long after the tasks were finished. Therefore we excluded the data points regarding these executions and re-charted the data. The corresponding control charts are shown in Fig. 6(b). This time, however, there appeared nine OCPs in the individuals chart and five OCPs in the moving range chart.

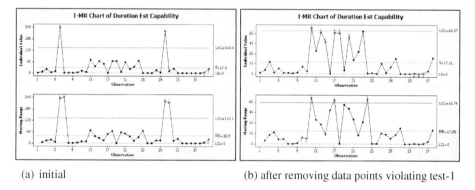

(a) initial         (b) after removing data points violating test-1

**Fig. 6.** XmR charts for the "duration estimation capability" measure of the task management process (combined clusters)

Considering the results of the process similarity assessment, we decided that these OCPs indicated a mixture of multiple-cause systems (data not being from the same source). We then charted the original data of process clusters C and D separately with respect to the duration estimation capability measure. The corresponding charts are shown in Fig. 7. For the process cluster C, there were one OCP in the individuals chart and two OCPs in the moving range chart. For the process cluster D, there were two OCP in the individuals chart and one OCPs in the moving range chart. An investigation for the reason of the OCP appearing in the individuals chart of the process cluster C showed again that in the executions that violated the test number 1, the closure of the task had been forgotten and carried out long after the tasks were finished. This investigation was carried out by completing a Process Execution

Questionnaire (PEQ) for the OCP via interviews with the task assigner and the task implementer. An example PEQ for the process execution 49 (regarding the OCP in the individuals chart in Fig. 7(a)) is given in Fig. 8.

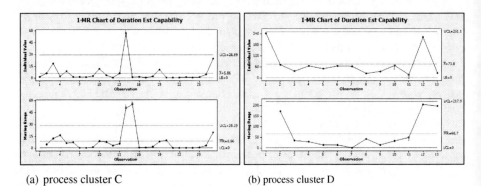

(a) process cluster C          (b) process cluster D

**Fig. 7.** XmR charts for the "duration estimation capability" measure of the task management process clusters C and D

After excluding the OCP regarding the process execution 49 from the data set, we re-charted the data of the process cluster C. The resulting charts are shown in Fig. 9(a). There appeared two OCPs in both the individuals and moving range charts. An investigation for the reasons of the OCPs appearing in the individuals chart showed that in the executions that violated the test number 1, the work plan changed but the task assignments were not updated on the change management tool. Therefore we removed these two points from the data set, and re-charted the data. The resulting control charts demonstrated controlled variation as shown in Fig. 9(b).

| Process Name: Task Management | | Recorded On: 12.June.2006 | |
|---|---|---|---|
| Process Execution No: 49 | | Recorded By: | |
| **External Attributes** | | Status (Yes/No) | Explanation |
| **PROCESS PERFORMERS** | | | |
| Q1 | Are process performers trained in their roles in the process? | Yes | |
| Q2 | Are process performers experienced in their roles in the process? | Yes | |
| Q3 | Are process performers differed per role basis during execution of the process? | No | |
| **PROCESS ENVIRONMENT** | | | |
| Q4 | Has there been a recent change in location? | No | |
| Q5 | Has there been a recent change in support systems? (infrastructure, technology, etc.) | No | |
| Q6 | Has there been a recent change in communication channels and mechanisms? (structure, media, etc.) | No | |
| Q7 | Has there been a recent change in funding and resources allocated for the process? | No | |
| Q8 | Has the process been tailored for this specific execution? | No | |
| **OTHER FACTORS (Please list if any)** | | | |
| | The task was forgotten to close. | | |

**Fig. 8.** An example PEQ completed for the task management process execution 49

On the other hand, for the OCPs appearing in the individuals chart of the process cluster D shown in Fig. 7(b), no assignable cause could be detected. Therefore, these points were considered as parts of the common execution of the processes and not removed from the data sets.

Assessment of Software Process and Metrics to Support Quantitative Understanding    117

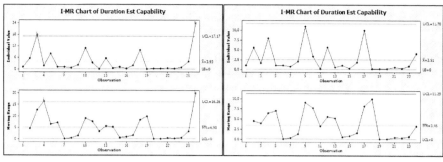

(a) after removing the process execution 49    (b) after removing the two OCPs in (a)

**Fig. 9.** XmR charts for the "duration estimation capability" measure of the task management process clusters C

When we looked at the XmR charts for the duration estimation capability measure of the process cluster C in Fig. 7(a) and Fig. 9(b), we observed that the mean values and the upper control limits in the charts in Fig. 9(b) decreased significantly. In the individuals charts, the mean value decreased to 2.51 from 5.86, and the upper control limit decreased to 11.70 from 28.89. Similarly, in the moving range charts, the mean value decreased to 3.45 from 8.66, and the upper control limit decreased to 11.29 from 28.29. These reductions indicated that the predictability of the task management process cluster C with respect to the duration estimation capability measure was improved due to a discrimination of the multiple cause systems in the task management process as a result of the process similarity assessment.

From the XmR charts for the duration estimation capability measure of the process cluster D in Fig. 7(b), we observed that the mean values and the upper control limits in the charts (X: mean 73.8, UCL 251.1; mR: mean 66.7, UCL 217.9) were very high when compared to those for the process cluster C. When we looked at the Process Similarity Matrix given in Fig. 2, we identified that the process cluster C differed from the process cluster D in only that its executions did not include the activity of the "verification of task implementation". Although this might be considered as a deficiency at the first glance, a review of the data showed that the actual finish date for the executions of the process cluster C was recorded at the time of finishing the task rather than at the time of its closure. Therefore, the values for the actual finish date recorded for the executions of the process cluster C were more dependable than those recorded for the executions of the process cluster D in representing the actual duration. Here we should remind that the usability of the data dependability attribute for the actual finish date measure was evaluated as "partial". The XmR charts for the duration estimation capability measure of the process cluster D validated this evaluation. The highness of the mean values and the upper control limits in the XmR charts in Fig. 7(b) was due to the late closure of the tasks by the task assigners.

As a result of the investigations of the assignable causes during the quantitative analysis, two primary reasons were detected for the OCPs that appeared in the individuals charts regarding the process cluster C: 1) Work plan changed and task assignment was not updated on the change management tool, and 2) Task closure was forgotten and performed later. We shared these findings with the project team. After the implementation, the project unit included reviews of task management data in regular progress monitoring to ensure task closures on time and to perform updates to task assignments as consistent with the project plan. In addition, normalizing the duration estimation capability measure with task size would provide more insight in the quantitative analysis; however, the task size had not been recorded in the process. Therefore, this was shared with the project team as another point of improvement.

## 4 Conclusions

Manufacturing disciplines have been applying quantitative techniques for the management of their processes for decades. The unique challenges of the software engineering discipline require the identification of practical processes, models, and guidelines on quality management. We developed a systematic and practical approach to evaluate the suitability of a software process and its measures for the quantitative analysis. This paper explained an implementation of the assessment approach for evaluating an undefined task management process and its duration estimation capability measure in a project unit of a government research agency. We utilized XmR charts in the quantitative analysis.

The assessment of process consistency indicated that the process data came from two different versions of the task management process in execution: The task management that included the activity of "task verification" and that did not. The assessment of measure usability indicated that the duration estimation capability measure was partially usable in the quantitative analysis. The quantitative analysis carried out by using the XmR charts showed that the task management process that did not include the activity of task verification demonstrated controlled variation, whereas the task management process that included the activity of task verification demonstrated uncontrolled variation, with respect to the duration estimation capability measure. These results were in keeping with the findings of the assessment approach.

The implementation facilitated an understanding of the task management process within a single project based on quantitative data. As a result of the implementation a number of weaknesses, including the lack of updates on the process data as consistent with the plan changes and the latency in the closure of the task assignments, were identified in the executions of the task management process. We also identified that the purpose of the quantitative analysis was important to relevantly cluster measurement data.

During the implementation, we observed that the act of measuring and analyzing the task management process was itself a vehicle for the understanding and improvement. We also observed that as long as process context and dynamics were understood process data could be utilized to serve a purpose. The existence of a well-defined approach to guide the quantitative analysis was an important motivator for the implementation. We hope this implementation motivates the software organizations in turn in assessing their processes and applying quantitative techniques to understand their potential for improvement.

# References

1. Weller, E.: Practical Applications of Statistical Process Control. IEEE Software 17(3), 48–55 (2000)
2. Florac, A.W., Carleton, A.D.: Statistical Process Control: Analyzing a Space Shuttle Onboard Software Process. IEEE Software 17(4), 97–106 (2000)
3. Baldassarre, M.T., Boffoli, N., Caivano, D., Visaggio, G.: Managing Software Process Improvement (SPI) through Statistical Process Control (SPC). In: Bomarius, F., Iida, H. (eds.) PROFES 2004. LNCS, vol. 3009, pp. 30–46. Springer, Heidelberg (2004)
4. Demirors, O., Sargut, K.U.: Utilization of Statistical Process Control (SPC) in Emergent Software Organizations: Pitfalls and Suggestions. Software Quality Journal 14(2), 135–157 (2006)
5. Card, D.N., Domzalski, K., Davies, G.: Making Statistics Part of Decision Making in an Engineering Organization. IEEE Software 25(3), 37–47 (2008)
6. Jalote, P., Dinesh, K., Raghavan, S., Bhashyam, M.R., Ramakrishnan, M.: Quantitative Quality Management through Defect Prediction and Statistical Process Control. In: 2nd World Quality Congress for Software, Yokohama, Japan (2000)
7. Pitterman, B.: Telcordia Technologies: The Journey to High Maturity. IEEE Sw. 17(4), 89–96 (2000)
8. McGarry, F., Decker, B.: Attaining Level 5 in CMM Process Maturity. IEEE Sw. 19(6), 87–96 (2002)
9. Jacob, L., Pillai, S.K.: Statistical Process Control to Improve Coding and Code Review. IEEE Software 20(3), 50–55 (2003)
10. Agrawal, M., Chari, K.: Software Effort, Quality, and Cycle Time: A Study of CMM Level 5 Projects. IEEE Transactions on Software Engineering 33(3), 145–156 (2007)
11. CMU/SEI-CMMI Product Team. CMMI for Development V1.3. CMU/SEI-2010-TR-033 (2010)
12. ISO/IEC. ISO/IEC 15504 Information Technology – Process Assessment, parts 1-7 (2003-2008)
13. Florac, A.W., Carleton, A.D.: Measuring the Software Process: Statistical Process Control for Software Process Improvement. Pearson Education, London (1999)
14. ISO/IEC. ISO/IEC 15939 Software Engineering – Software Measurement Process (2002)
15. McGarry, J., Card, D., Jones, C., Layman, B., Clark, E., Dean, J., Hall, F.: Practical Software Measurement: Objective Information for Decision Makers, 1st edn. Addison-Wesley, Reading (2001)
16. Dyba, T., Kampenes, V.B., Sjoberg, D.I.K.: A systematic review of statistical power in software engineering experiments. Information and Software Technology 48(8), 745–755 (2006)
17. Chang, C., Chu, C.: Improvement of causal analysis using multivariate statistical process control. Software Quality Control 16(3), 377–409 (2008)
18. Jalote, P., Saxena, A.: Optimum Control Limits for Employing Statistical Process Control in Software Process. IEEE Transactions on Software Engineering 28(12), 1126–1134 (2002)
19. Lantzy, M.A.: Application of Statistical Process Control to Software Processes. In: 9th Washington Ada Symposium on Empowering Software Users and Developers, pp. 113–123 (1992)
20. Card, D.: Statistical Process Control for Software? IEEE Software 11(3), 95–97 (1994)
21. Radice, R.: Statistical Process Control for Software Projects. In: 10th Software Engineering Process Group Conference, Chicago, Illinois (1998)

22. Paulk, M.C., Goldenson, D., White, D.M.: The 1999 Survey of High Maturity Organizations. Special Report, CMU/SEI-2000-SR-002 (2000)
23. Paulk, M.C., Chrissis, M.B.: The November 1999 High Maturity Workshop. Special Report, CMU/SEI-2000-SR-003 (2000)
24. Paulk, M.C., Chrissis, M.B.: The 2001 High Maturity Workshop. Special Report. CMU/SEI-2001-SR-014 (2000)
25. Kasunic, M.: The State of Software Measurement Practice: Results of 2006 Survey. Technical Report, CMU/SEI-2006-TR-009 (2006)
26. Stoddard II, R.W., Goldenson, D.R., Zubrow, D., Harper, E.: CMMI High Maturity Measurement and Analysis Workshop Report: March 2008. Technical Note, CMU/SEI-2008-TN-027 (2008)
27. Tarhan, A., Demirors, O.: Investigating the Effect of Variations in Test Development Process: A Case from a Safety-Critical System. Software Quality Journal (in press)
28. Tarhan, A., Demirors, O.: Assessment of Software Process and Metrics for Quantitative Understanding. In: Cuadrado-Gallego, J.J., Braungarten, R., Dumke, R.R., Abran, A. (eds.) IWSM-Mensura 2007. LNCS, vol. 4895, pp. 102–113. Springer, Heidelberg (2008)
29. Kirbas, S., Tarhan, A., Demirors, O.: An Assessment and Analysis Tool for Statistical Process Control of Software Processes. In: SPICE 2007 Conference, Seoul, Korea (2007)
30. Wheeler, D.J.: Advanced Topics in Statistical Process Control. SPC Press (1995)
31. ISO. ISO 9001, Quality Management System – Requirements (2008)
32. MINITAB Statistical Software (Release 14), http://www.minitab.com/

# Methodical Enhancement of Maturity Level: "SPICE" and "SixSigma" Intertwine

Timo Karasch and Jens Peter Benthaus

IAV GmbH, Nordhoffstrasse 5, 38518 Gifhorn, Germany
{timo.karasch,jens.peter.benthaus}@iav.de

**Abstract.** That there is a direct correlation between SixSigma, a method for process improvement, and SPICE is already cognizable by the name. While the norm describes, which optimizations should be implemented, SixSigma offers a set of methods therefore. How an adequate support of the continuously enhancement of the maturity level could be guaranteed even for lower levels, is shown in this contribution. Based on experiences of some Green-Belt projects, the key-questions are: 1) How can I use SixSigma methods for process improvement in SPICE? 2) What support can these methods offer for the maturity levels 1 to 5? 3) What benefits brings SixSigma in this correlation? And 4) What risks shall be faced?

**Keywords:** SixSigma, Enhancement of Maturity Level, Methods for Process Improvement, Experience Report.

## 1 At First a Short Mind Game

Imagine a team of eleven players. All are running across a lawn in an uncoordinated way. They do not know anything about their reason of being there or their function. They do not even notice the further eleven people, their opponents, beside them. By a fluke one of them finds a small round ball made of leather and tries to handle it. As the ball jumps off his foot, it spins to the other side of the field and reaches a net.

Congratulations, your new soccer team "Organisation Level 5" has just won its first match of its young carrier, or rather finished the project successfully. Unfortunately nobody can explain the success, much less repeat it. Now it is your challenge to lead the team to the "Champions-League", that is to create a "SPICE Level 5"-team. But do not panic, your new assistant "SixSigma" will help you.

## 2 Problem

Since a few years SPICE is "state of the art". But while the standard defines requirements for process improvement, it gives no methods. To achieve a sustainable enhancement of maturity level, there are still some gaps. We will try to group them into three categories (table 1).

**Table 1.** Categories of SPICE gaps

| What should be improved? | For our mind game: Where should I start improvement in my team "Organisation Level 5"? It is not helpful to show the players complex tactical moves while they still have problems with passing. |
|---|---|
| How should it be improved? | How can I help my players to learn their lessons? If I yell at them for every bad pass but they are passing the next ball to the opponent again, perhaps my method is inappropriate. |
| How good is the improvement? | Have my actions been successful? The final score of our next match is a first indicator. In addition we can for example analyse the precision of passing and carry out a video analysis of the match together with the team. |

## 3 Problem Solver

Now we will see, what our new assistant "SixSigma" can provide. To keep our example: We will not explain the whole CV in this report. SixSigma is a lot more than just tools. An important reason for the usage in improvement projects is the fact that achieved optimizations can be shown as earned savings. But in this case we will concentrate on the "toolbox", "metrics" and "methods".

*Toolbox:* Including proven tools for the statistical analysis, otherwise mainly used in development and design and for visualisation of sequences of actions and processes. For our mind game: Our assistant can display the possession of the ball versus the course of the match, show the effect of training methods and visualise tactical moves.

*Metrics:* These are mainly statistical measurements showing the current status. A target status shall be defined for the improvements. For our mind game: Our assistant does not only know the result of our game, but also statistics like "won and lost duels" during the whole match.

*Methods:* These procedures are proven and can be used as a roadmap for the execution of SixSigma projects. For our mind game: Our assistant can not only train simple units. He can design, plan, adjust and execute whole training programs and ensure their sustainability.

## 4 Approach

These key features of our problem solver can be used in several situations for several times. We will now just give a few examples. For each maturity level the main challenges are listed and a way to solve them by using SixSigma methods is shown.

By using the support of our approach we will follow our team "Organisation Level 5" stepwise across the respective maturity levels. As shown in the introduction, our initial position will be Level 0 which is characterized by its missing repeatability and reproducibility. Our team consists of single heroes, using their time for fire fighting. So far, they all do not have enough time for improvements.

## 4.1 Level 1

*Problem:* For most of the members, a deeper understanding of processes is missing. The employees do not know or are not aware of the sequence of operations. It is not even possible, to see the guidelines of SPICE in a simple coherence.

*Example for solution:* SIPOC

In a SIPOC for example all base practices of a process can be displayed. In a next step, all work products are allocated as outputs and their receivers as customers. On the left side all work products necessary for a base practice are allocated as inputs. The responsible persons are shown as suppliers.

The diagram helps members to understand the coherence of this process. In addition, it provides an overview of work products and interfaces. Thereby, it is a useful extension for the process description in the PAM.

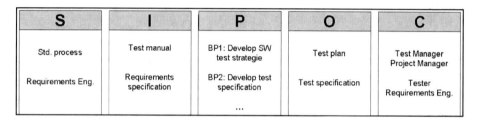

**Fig. 1.** Example "SIPOC"

*Problem:* Often influences on single activities or work products are not known. Therefore, influencing variables are underestimated or even ignored.

*Example for solution:* Ishikawa (Fishbone diagram)

The diagram shows all influences on the activity "test preparation". These do imply materials like the software that is to be tested and the test specification. As manpower the necessary resources like the tester himself is mentioned. Test equipment like the test configuration and infrastructure belongs to the machinery. The management delivers the planning and the strategy for the test. The diagram could be completed with all influences. Afterwards, it can be used as a checklist for test preparation. In addition, it is possible to analyse, where process coordination is necessary and can serve as a basis for the identification of critical influencing variables as shown in the next example.

*Problem:* Even if effects are known it is often not identified which of them have critical influence on the problem, activity or work product.

*Example for solution:* Pareto chart

The usage of a pareto chart can help you getting a quick overview of the main impacts. For example the reasons for test failures.

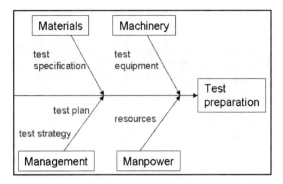

**Fig. 2.** Ishikawa Diagram

There are even more utilities that help a team member achieving a Level 1. But back to our short mind game: We have presented the objectives and main elements of the game to our players. They know now the influences whether it will be a win or a defeat and why they were beaten in the past. So they understand that they all are soccer players and winning means, that they score more goals than the opponent. Above all they know the rules of the game and recognised that there are further players on the field.

This was the first step. Our team now consists of eleven individualists. They do a good job, but they act still uncoordinated. Like in the youngest junior-league all players will try to get the ball and score a goal. It is time to create a team, time to manage our players. This leads us to the next level.

### 4.2 Level 2

*Problem:* Every intention starts with the collection of requirements and hence, in terms of project management with the analysis of all relevant stakeholders. The list of stakeholders is often missing and requirements are incomplete and not or rarely aligned and estimated.

*Example for solution:* Stakeholder analysis / VOC, VOB

One part of SixSigma is a complete stakeholder analysis. The requirements, they claim to the intention are collected as "Voice of customer" and "Voice of business" like already known from development projects. They shall be aligned, because they serve as the basis for every prioritization of actions during the project. Improvements which support the VOC create customer satisfaction (e.g. 0-defect quality). The provision for VOB helps designing processes more effective and efficient (e.g. minimise the time for storage).

*Problem:* Actions for improvement are often selected intuitively. But an analysis of the necessary effort and the existing dependencies is mandatory.

*Example for solution:* Decision matrix

The diagram below shows the main work products of the five software processes. Considering the dependencies of work products and processes you can see for example that an improvement of the requirement specification and test specification affects all other processes. This helps a project team, to implement actions not just in one process, but crossing the limits of processes to decrease the necessary resources und improve the suitability of the solution. You can even identify actions outside of the projects. Problems can be solved together. A faster solution can be found, experience is shared and a basis for a level 3 is provided.

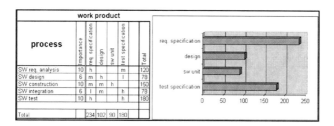

**Fig. 3.** Example "Decision matrix"

*Problem:* What is not defined can not be measured. Often measurable objectives to control the project are missing.

*Example for solution:* Metrics (DPMO, σ-level)

Metrics are already useful for a Level 2. They set measurements for the selected processes and allow the controlling by the project management. After defining an upper limit, actions are implemented to decrease e.g. the DPMO sustainably and measurable.

Further utilities, to support the project team achieving a Level 2 can be found. With help of our assistant "SixSigma" we have formed a team out of our eleven players. Everybody is allocated to his functions. We now have a goal keeper, defence, midfield and a centre. We recognized that for example even the fans have a strong effect on our success and we started to define measureable goals for our next matches. By analysing the requirements of our club administration, the experience with competitors in the league and additional stakeholders, we identified ideal training methods for our team. We now have a managed team that can perform together.

But we also discover that all these analysing and planning costs a lot of resources. Team formations, training schedules etc. will always recur in a similar way. So we should try to standardise them. We move towards the next level.

## 4.3 Level 3

*Problem:* Standard processes shall be visualised the way that the expected readers can understand and implement them easily.

*Example for solution:* Process mapping (e.g. Swim lane diagrams)

The first step is to translate the sequence of operations in the mind of the members into a process diagram. One utility, the SIPOC, we already know. As processes should be usable in various projects we have to consider that members will need different levels of detail. Project members who already know their process very well just need a quick overview like a flow chart. If people want to know what activities and work products are connected with an activity you will need the chance to zoom in the process. This could be done for example by a swim lane diagram shown aside. The next level of detail gives the novices to the process examples and concrete utilities for the execution of the activities. This could be done by using additional process manuals and catalogues of examples.

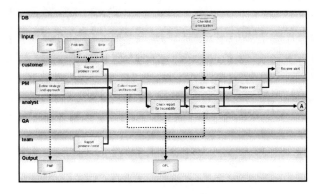

**Fig. 4.** Example of a Swim lane diagram

*Problem:* Experiences of the deployment of standard processes are rarely used in later projects or for the improvement of processes and templates.

*Example for solution:* Lessons learnt workshop

This technique is known from many project management methods as well. It is a mandatory element in the last phase of SixSigma projects. Often a special agenda (like the example aside) or checklist for performing a lessons learnt workshop already exists.

*Problem:* Most processes are designed or optimized without considering specific basic conditions. An evaluation of changes against the actual project risks is missing.

*Example for solution:* Risk analysis (e.g. FMEA)

An FMEA can be used for various challenges, but the preparation requires a lot of resources. You can use it for example to collect all activities (base practices) of the process combined with possible failures and their severity. Actions for process improvement should start, where a risk with a high RPN (risk priority number) is identified.

To support an organisation in achieving Level 3, some further utilities can be used as well. We now have defined standard team formations for our team "Organisation

Level 5" that can be adjusted to the conditions of our next opponent. All functions in our team are described. Formations, training units, etc. get improved using the feedback of our players. The risks for our operation are identified. For example we discovered that the combination of some training units and a tough playing schedule increases the risk of injuries of our players. That is why the training schedule was adjusted.

As we eliminated the factor "accident" in our team, "Organisation Level 5" starts to perform and improve permanently. What still keep us apart of our big aim are external influences on our performance. As we want to control them as well we are heading for the next level.

## 4.4 Level 4

*Problem:* The quality and especially the stability of a process is most of the time based on the "gut feeling" of the responsible person, not on data.

*Example for solution:* Mean, Deviation and Control limits

You use diagrams of the chronological sequence of measurements for further analysis. An important part is the mean of the over all data that gives you a prediction for further measurements. For the DPMO example from Level 2 it will show you the expected failures per lines of code. In terms of a normal distribution all results will fit in a bell-shaped curve. The percent value shows you the probability, the result will be in this area. Using the σ-value, the upper and lower control limits can be calculated. As all measurements lie within these limits our process is stable and you can predict the number of failures per lines of code for example. Data outside these limits will be handled in the next example for solution.

*Problem:* SPICE expects for Level 4 actions to reduce "out-of-control"-values. How this can be implemented is not mentioned.

*Example for solution:* SPC chart (special causes)

The chart shows the chronological sequence of our process. The example above shows the upper and lower control limits (UCL and LCL). Values that are "out-of-control" and that should be analysed for their cause (for example in a 5-why analysis) are all measurement data lying outside the control limits.

*Problem:* We already started with the measurement of metrics. But the characteristics of measurement systems are still unknown so is the nature of the metrics itself.

*Example for solution:* R-charts and Regression analysis

As we collect data for process control, we use measurement systems which are creating failures by themselves. A measurement system analysis (MSA) will help to evaluate the repeatability and reproducibility of the collected data.

We use this data for combinations of measurements according to our assumptions. Regression analysis can help to check, whether our assumptions are correct and what kind of dependency exists between two ore more inputs.

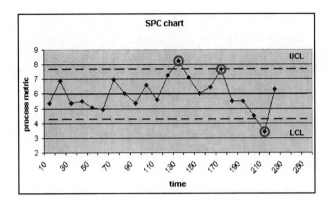

**Fig. 5.** Example "SPC chart" with special causes

These are only the most important analysis methods for support of a Level 4. There will be many more. Just for the estimation of necessary actions for example a couple of methods and metrics of SixSigma will be used.

To identify further improvements for our soccer team we for example analyse whether there is a correlation between the duration of the match and the won and lost duels. We can even search for influences of the temperature, date and time or the height above sea level on our team performance. For our metrics "pass completion", "won duels" etc. exist predictions with a specific probability. We can identify "special causes" and analyse them. These are for example injuries, changes of weather or even wrong decisions of the referee. In some cases actions were implemented to decrease the influence of these causes, like an optimization of the training units for different weather conditions or briefings with the players to prepare them for the referee.

So our team is successful and can cope with setbacks. If this would be already everything every team would be able to reach the top of the league. But some teams are still not "good enough". So we have to increase our performance and reduce the variation of our results. We go about the last level.

### 4.5 Level 5

*Problem:* The compliance with customer requirements is not only up to the elimination of "out-of-control"-values. Whether a process itself is suitable to fulfil these requirements is still unknown.

*Example for solution:* Specification limits

In the level above we have seen that our process can be illustrated as a normal distribution within the control limits. The customer requirements will now define the upper and lower specification limits (USL and LSL). If the mean is located outside these limits our process will be totally unsuitable to fulfil these requirements. Overlaps the variation (within the control limits) one of the specification limits, we are producing scrap in the production or rather failures in the development. In both cases an optimization of the process is necessary to improve the performance and reduce the variation.

*Problem:* Optimization of the process should be done regarding to the requirements of all stakeholders. This is for example our management. An orientation to our business goals is demanded, but the implementation is not described in SPICE.

*Example for solution:* QFD

The QFD (also known as: house of quality) can help to break down general goals of the organisation to concrete requirements for our processes and prioritize them at the same time. The example below shows, how the mission of the organisation is rated and prioritized against the vision. We can analyse the goals afterwards. If we focus on our process improvement intention we will be able to weight the necessary requirements against these goals the same way. A deviation of the process inputs shows us the parameters that are critical to quality (CTQ). The QFD also gives you the chance to rate the dependencies between the factors.

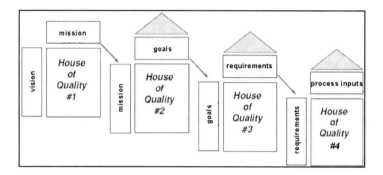

**Fig. 6.** Example "QFD"

*Problem:* In a final step we need assurance, that all optimizations were really able to improve our performance and reduce the variation.

*Example for solution:* SPC chart (common causes)

We will use the SPC chart again to control the impact of our changes to the process and to check their effectiveness. An improvement of the performance should be observable by a shift of the mean in the specific direction for the new measurement data. If we were able to reduce common causes furthermore, the σ-value would shrink. This is simultaneously an increasing of the σ-level. It is observable by a decreasing of the distance of the control limits. Our prediction gets a higher precision.

We were able to identify further actions for optimization with the help of our assistant "SixSigma". Our club made new investments in innovative equipment. Advanced training methods are rolled out and the process-related conditions for our aim, the achievement of the "Champions-League", are accomplished. Our team "Organisation Level 5" just arrived.

## 5 Benefits

Summarizing our approach, we discover the benefits of the combination of the requirements of SPICE and the key features of SixSigma.

The basis for any improvement is acceptance by all members. This is supported by SixSigma, as it helps to fix the intuition for quality in the organisation, raises the awareness of problems and deepens the understanding for processes. Together this creates established process thinking which is the basis for SixSigma's key features:

- The across all levels introduced toolbox will help to identify, WHAT should be improved.
- The methods (roadmaps) will give us a clue, HOW an improvement project should be planned and controlled.
- Finally the metrics can be used to determine HOW GOOD our improvements really are.

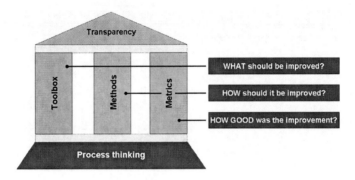

**Fig. 7.** Benefits of this approach

The roof of our combination is process transparency, meaning: The tools will show all prioritizations and decisions for everybody in a traceable way. The metrics show a success story and the methods can guarantee a continuous controlling and cooperation between all improvement intentions. That is the way every member can participate in results, achievements and experience of all improvement projects.

## 6 Risks

To really deploy the complete benefits some risks have to be faced. Because of the complexity and versatility of the SixSigma contents some side effects that can appear you have to prevent early. The most important problems SixSigma can come up with while achieving the specific maturity levels are shown in the table below:

Methodical Enhancement of Maturity Level: "SPICE" and "SixSigma" Intertwine    131

**Table 2.** Risks

| | | |
|---|---|---|
| Level 1 | heroism | Because every member will get a tool to face problems himself, the heroism of a level 0 can be enforced. That is why problem solutions should always progress to the real cause, to gain sustainable results. A consistent consolidation of all improvements and results is mandatory! |
| | too many loops | Every analysis gives us the chance to get deeper into the problems, but this can cause too many loops. The 80/20-rule (Pareto) should be followed. |
| | control of members | SixSigma works primarily with metrics. But these can leave the impression, they were applied to the members, what would be a control mechanism. It must be necessarily guaranteed, that metrics are measured process-related, not people-related. |
| Level 2 | statistics incorrect / unsuitable | Churchill said: „the only statistics you can trust are those you falsified yourself". The suitability of every statistic must be checked und their results must be analysed critically, before you draw any conclusions. |
| | duplicated solutions | Even a consistent procedure for process improvements cannot guarantee that solutions for identical problems are identified in different projects. A comprehensive controlling of all intentions can be helpful. |
| | team in team | In process improvement teams a certain kind of dynamic starts to grow. Its members must be integrated in the development team as well so that no new elite, a team in the team, is founded and solutions have a certain correspondence to the project. |
| Level 3 | over-engineering | With help of the introduced tools, processes can be designed in the minutest details. This can lead to an over-engineering of these documentations. Processes shall still be lean and easy to implement. Use deeper details only where they are necessary e.g. for critical activities. |
| | no time for stabilisation | Especially at the beginning of improvement projects a lot of ideas for actions are identified. Every change should grow until it is established and experience can be gained. Wait for a stabilisation of the new process until you start new actions for improvements. |
| | unsuitable process selection | A redesign or an improvement is mostly done for processes that are already used and well known. But you should always check new processes for their suitability as well. Besides you should rate the existing processes with a critical view and bear the consequences. |

**Table 2.** (*continued*)

| | | |
|---|---|---|
| Level 4 | unsuitable metrics | Not every correlation is meaningful. At first gain sufficient experience in metrics before you rely on their validity. |
| | measuring inaccuracy | Especially in development projects many process measurements can only be estimated. For example the duration of some processes overlap. You will always have to check how precise your data really is. |
| | reaction without cause | A lot of actions are started, because an "out-of-control"-value is suspected. Analyse, whether this is really an "out-of-control"-value (special cause) or just the normal variation of your process (common cause). Otherwise the problem could get even worse. |
| Level 5 | innovations vs. requirements | Try to find a compromise between possible innovations (based on business goals) and the requirements of your customer. In many cases they will conflict. |
| | improvements instead of innovation | Do not spare too many resources for improvement. Check, whether a "repairing" of the process is profitable or an innovation should be developed. |

## 7 Conclusion

Summing up you should always keep an eye on the following topics:

- Try to build the basis for SixSigma early. All members will be able to gain experience and you will deepen the process thinking.
- Use the combination of the requirements of SPICE and the key features of SixSigma as you establish a consistent controlling of all improvement activities.
- Celebrate your success. Use results and metrics to present improvements and savings to all; the members and the management.

At last: Take your time. With an effort we lead our soccer team "Organisation Level 5" from the initial position at the bottom to the "Champions-League". Be aware of being relegated because you pushed too hard at the instant. Provide enough time for the three topics above, and your soccer team may perhaps even become the winner of the "Champions-League".

## References

1. Hoermann, K., Dittmann, L., Hindel, B., Mueller, M.: SPICE in der Praxis. dpunkt.verlag GmbH (2006)
2. Pande, P.S., Holpp, L.: What is six sigma?, 1st edn. McGraw-Hill, New York (2001)

# Organizational Support for Process Improvement – Results of an International Survey

Marion Lepmets, Eric Ras, and Alain Renault

Public Research Centre Henri Tudor,
29 J.F.Kennedy ave., Luxembourg, Luxembourg
{Marion.Lepmets,Eric.Ras,Alain.Renault}@Tudor.lu

**Abstract.** Organizational support for process improvement initiatives is vital to the success of these improvements. This paper describes and explores the steps of supporting process improvements described in the new international standard ISO/IEC 33014 that is currently being developed.

**Keywords:** process improvement, organizational support, organizational readiness, ISO/IEC 33014.

## 1 Introduction

A lot of studies have been conducted on process improvement in the last 15-20 years, particularly in software engineering domain. There are numerous case studies about the success and key success factors of process improvement [1-7]. One of the most often mentioned key success factors is the organizational support for process improvement initiative i.e., what is the organizational capability for change and for taking on an improvement effort [8]. According to Korsaa et al. [9] over 70% of all process improvement initiatives fail mainly because of the lack of management commitment and the poor understanding of competencies, roles, responsibilities of process improvement activities, and tasks. In order to help organizations to better support their process improvements, a new international standard ISO/IEC 33014 is currently being developed for that purpose.

ISO/IEC 33014 is a guide for process improvement focusing on how continual process improvement can be enhanced based on the necessary support organizations could provide to the improvement initiative. According to ISO/IEC 33014 [10], process improvement programmes or improvement projects are the work people do to realize the change. Continual process improvement is a cycle based upon the premise that in order to always meet customer needs, organizations must continuously improve [10]. ISO/IEC 33014 is largely based on the ideas described in Improve IT [11] where enhancing organizational support for process improvement is called improvability i.e., improving the organization's ability to improve. Statz [8] calls similar aspects of support as organizational readiness for process improvement.

Organizational readiness for process improvement according to Statz is addressed by two measurable concepts: alignment and commitment, and process improvement capability. The alignment and commitment category seeks to determine whether or not the organization is committed to the process improvement project with sufficient involvement of management and availability of resources to enable the process improvement project to be successful. The process improvement capability measures the overall organizational capability to do process improvement, for making organizational changes and for establishing current process capability baselines.

ISO/IEC 33014 is twofold – it describes how organizations can support process improvements, and how they can enhance this support to increase the success rate for process improvements.

In this paper we focus on the availability of the organizational support and we are not addressing the improvement of these organizational support steps, the latter being called improvability in ISO/IEC 33014. Organizational support for process improvement according to ISO/IEC 33014 describes measures on three organizational levels – operational, tactical and strategic levels. The improvement on operational level is the improvement of processes that is described in greater detail in ISO/IEC 15504-2 [12]. The improvement on tactical and strategic levels could be viewed as the organizational support for the process improvement on operational level. It is on the tactical and strategic levels that most organizations face difficulties that might prevent them from succeeding in process improvement.

The aim of this study is to discover the conditions in which organizations prepare for and support process improvements. We will first describe the steps that organizations can take to better support process improvement as illustrated in ISO/IEC 33014. We will then describe our research questions in greater detail and present the industry survey used for data collection and the sample. Finally, we will illustrate the conditions in which organizations are preparing themselves and supporting process improvement initiatives based on the received data.

## 2  Designing the Survey Questionnaire Based on ISO/IEC 33014

According to ISO/IEC 33014 [10] there are steps organizations can take to better support process improvement on both the strategic and tactical levels. The following list provides four steps on the strategic and three on the tactical level.

Steps that provide *strategic* support for process improvement i.e., support from the entire organization to ensure process improvement success:

1. *Identifying and communicating organization's business goals* – business goals are the drivers for visions, strategies, decisions, and many fundamental elements that support improvements. The objectives for improvement should be set based on an analysis of the organization's business goals and existing stimuli for improvement. The more clearly the business goals link to objectives of improvement, the higher the probability of success for the improvement programme;
2. *Identify the scope of organizational change* – to ensure the best possible setup of a process improvement project (or program) it is important to clarify the situation, the scope, and the vision for change at organizational level;

3. *Decide upon the overall change strategy for the organization* – changing an organization usually means selecting a change strategy from among many available change models. To ensure the right understanding of the situation of the organization and the most suitable model and strategy, this decision has to be done together with the management;
4. *Get management's support and commitment for the improvement* – it is essential to build executive awareness of the necessity for a process improvement programme, which requires both managerial and financial commitments. According to Statz (2005), organizational commitment and management involvement to process improvement enable the improvement programme or initiative to be successful;

Steps that provide *tactical* support for process improvement i.e., how to identify what and how to improve and who is doing what to ensure smooth and successful process improvement:

5. *Identify the process improvement goals and scope* – it is necessary to analyse organization's business goals against the improvement initiative to ensure the foundation of the initiative is in budget, and identify the main process improvement priorities and their relation to the organization's improvement strategy;
6. *Allocate the roles and responsibilities for the improvement in the organization* – successful change is highly dependent on the way improvement work is organized in an organization. Improvement work should aim to enhance communication, keep improvers bound to practice, and deploy the improvements in the organization;
7. *Conduct a process assessment* – process assessment results describe what is the current capability of the processes and leads to the operational level of process improvement as an input to the developing of process improvement action plan;

## 2.1 Research Goals

In the last 20 years, various process models and guides have been developed for conducting process assessment like CMMI and ISO/IEC 15504. There are also models describing the steps of process improvement in IDEAL [13] and in ISO/IEC 15504 [12]. Too little has been said about how to prepare your organization for process improvement and increase your chances to succeed in this timely and resource-consuming initiative. ISO/IEC 33014 describes the support elements on the strategic and tactical level that help carry out process improvements. It also describes how to enhance these organizational support elements that will, in return, increase the chances to manage successful continual process improvements.

This study aims to discover the conditions in which organizations prepare for and support process improvements. An international online survey was conducted to gather data about process improvement readiness and support during two months of early 2011. Since survey cases are typically used for establishing proof or verifying propositions [14], following are the propositions in the form of research questions.

There are various ways to improve processes – we can improve the way we work as a result of having better technical tools at hand or by training ourselves. When we work alone on our personal process improvements, we do not need the same level of support from the organization that is required for successful organization-wide process improvement initiatives. Process improvement initiatives are means to develop an organization's processes to more effectively meet its business goals. Process assessments are used to find out the capability of the process to reach this goal [15]. Process assessment should revisit and communicate the organization's business goals and align the goals of process improvement and the prioritization of the processes for improvement to organization's business goals. Since process assessment is a time-consuming and resource-demanding undertaking mostly conducted with the aim to later improve the processes, we believe that *organizations that conduct standard or model based process assessments make a bigger effort to have organizational readiness and support for process improvement.*

Although ISO/IEC 33014 claims that organizations using ISO 9000 are more likely to prepare for and support process improvements, *we believe that organizational readiness for process improvement is not related to any one specific model they use.* Implementing any organization-wide model, method or framework requires similar organizational support, and organizations with the prior experience of implementing them are more prepared to face similar challenges when implementing process improvement initiatives.

Process improvement is timely and resource-demanding and decreasing the possibility of failure is vital to any organization undertaking these initiatives. We support the claim made in 33014 that *all organizations improving their processes should prepare for and support the improvements regardless of their size and core business area.*

### 2.2 Data Collection and Description of the Sample

An online survey was used to collect data from industry about the readiness and support that organizations provide to process improvement initiatives. Since only the organizations interested and/or experienced in process improvement were targeted, the snowball sampling technique of the non-probability sampling method was used in this research. The request for distributing and responding to the survey was sent to various working groups in ISO (International Standardization Organization) subcommittee 7 (ISO/IEC SC7), to companies, process improvement consultants, researchers, and non-profit organizations promoting process improvement worldwide.

After two months, there were 50 completed responses received. Out of the 50 responses, the distribution between software development and IT service providing organizations was almost equal forming more than 50% of all the responses.

Ten responses out of 50 came from organizations providing IT services, nine from software development organizations, and another ten from organizations both developing software and providing IT services. There was one response from an organization that did not categorize into any given business area. (Fig. 1).

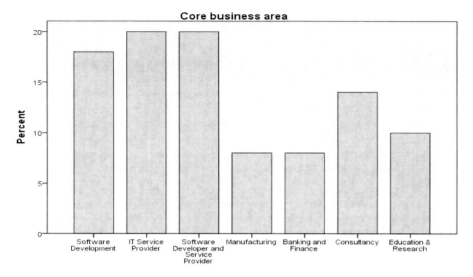

**Fig. 1.** Core business area of respondents' organizations

Over half of the responses came from large organizations employing more than 250 employees (62%), 18% from medium-sized organizations employing 50 to 249 employees and 10% from both small (with 9 to 50 employees) and micro (up to 10 employees) organizations (Fig. 2).

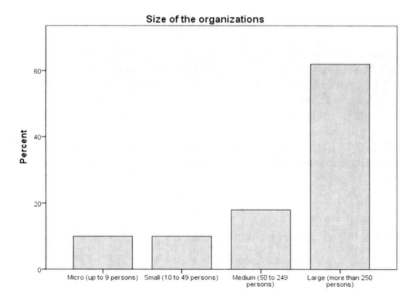

**Fig. 2.** Size of respondents' organizations

Since cultural aspects play an important role in process improvement initiatives, we also sought information about the location of the respondents' organizational headquarters. The responses have the following geographical distribution: 68% of responses came from Europe with Finland being the most active respondent, 16% of responses came from USA, 6% from Canada, 4% from Mexico and Australia, and 2% from Peru. Despite the fact that India is heavily using process models, no responses were unfortunately received from India.

In order for us to understand the conditions in which companies support process improvements more, we also sought information about the standards and frameworks that the organizations implemented. Table 1 illustrates the responses about the standards, models, and frameworks that were used in the respondents' organizations.

**Table 1.** Standards, methods and frameworks used (n=50)

| Own experience and knowledge | PSP/TSP | ISO/IEC 15504 | CMMI | Six Sigma | ITIL | CoBIT | ISO/IEC 20000 | ISO 9000 | Lean | Theory of Constraints | No improvement methods used |
|---|---|---|---|---|---|---|---|---|---|---|---|
| 33 | 6 | 15 | 21 | 12 | 24 | 7 | 6 | 17 | 18 | 5 | 1 |

The survey allowed the respondents to choose as many responses as were relevant in their case. The majority of organizations use their own knowledge and experience in process improvement, accompanied closely with CMMI, Lean and ISO 9000.

As Jones [16] points out, it is not wise to start process improvement if managers do not calculate the return on investment or collect data to demonstrate the progress. There are various ways to measure the progress and Table 2 illustrates how process improvements were measured among the respondents' organizations. In almost half of the cases, process improvements were measured based on the customer and stakeholder satisfaction, and by measuring personal performance and productivity. Standard or process model based assessments indicate the strengths and weaknesses in processes and suggest how to improve them. Standard or model based process assessments were carried out in 19 cases out of the overall 50 responses.

**Table 2.** Measuring process improvements (n=50)

| Improvements are not measured | Measuring personal performance and/or productivity | Evaluating the achievement of product or service quality requirements | Evaluating the achievement of project or service performance objectives | Measuring project productivity | Conducting model/standard based process assessments | Evaluating stakeholder/customer satisfaction | Measuring organizational productivity | Evaluating the achievement of organizational goals | Calculating the return on investment to process improvement |
|---|---|---|---|---|---|---|---|---|---|
| 4 | 23 | 20 | 17 | 18 | 19 | 25 | 13 | 23 | 5 |

## 2.3 Data Analyses

We asked the respondents to indicate if they have taken the steps described in ISO/IEC 33014 on the strategic and tactical levels to support process improvements and increase their chances to succeed in improving processes. Table 3 illustrates the steps supporting processes improvements where the first four are on the tactical level and the last four are on the strategic level. As a positive surprise, the goals for process improvements are set almost in all organizations and setting the scope for process improvements follows closely. In half of the cases, the companies conduct process assessments to understand their current situation and determine the necessary improvements. Also roles and responsibilities for process improvement work are allocated in 60% of the cases. All these elements describe the process improvement support organizations provide on the tactical level.

Steps to support process improvement are not as often taken on the strategic level as they are on the tactical level. Management's support is sought in 62% of the cases, but awareness of the organization's business goals, the change strategy and the scope of change are present in less than half of the responses.

**Table 3.** Organizational readiness and support to process improvement

| | Organizational support on strategic and tactical levels | Count | Percentage |
|---|---|---|---|
| *tactical level* | Identify the process improvement goals | 41 | 82% |
| | Identify the process improvement scope | 38 | 76% |
| | Allocate the roles and responsibilities for the improvements in the organization | 30 | 60% |
| | Conduct a process assessment | 25 | 50% |
| *strategic level* | Get management's support and commitment for the improvement | 31 | 62% |
| | Decide upon the change strategy for the organization | 22 | 44% |
| | Set the scope of change in the organization | 20 | 40% |
| | Identify and communicate the organization's business goals to the staff | 23 | 46% |

To be able to understand if the background characteristics of the organization play a role in the organizational support for process improvements, we looked at the size and the core business area of the respondents' organizations to see whether some are more prone to organizational readiness than the others. Following table (Table 4) illustrates the organizations' core business areas and the tactical and strategic support for process improvements. Regardless of the business area, organizations' tactical level support for process improvement is more apparent than that on the strategic level. Setting the scope of the organizational change and the awareness of organization's business goals was equally low across the core business areas with the exception of the manufacturing organizations. Surprisingly, even process improvement consultants did not know the organization's business goals in over half of the cases. This, in turn, might lead to process improvement goals not being aligned to the organization's business goals.

**Table 4.** supporting process improvement by organization's core business area

|  |  | Software Development | IT Service Provider | Software Developer&Service Provider | Manufacturing | Banking and Finance. | Consultancy | Other | Education & Research | TOTAL |
|---|---|---|---|---|---|---|---|---|---|---|
|  | n | 9 | 10 | 10 | 4 | 4 | 7 | 1 | 5 | 50 |
| *tactical level* | Identify the process improvement goals | 7 | 8 | 10 | 4 | 1 | 6 | 1 | 4 | 41 |
|  | Identify the process improvement scope | 7 | 7 | 10 | 1 | 1 | 7 | 1 | 4 | 38 |
|  | Allocate the roles and responsibilities for the improvements | 8 | 6 | 4 | 2 | 2 | 6 | 1 | 1 | 30 |
|  | Conducting process assessment | 4 | 5 | 7 | 1 | 1 | 4 | 1 | 2 | 25 |
| *strategic level* | Get management's support and commitment for the improvement | 7 | 5 | 6 | 2 | 3 | 5 | 1 | 2 | 31 |
|  | Decide upon the change strategy for the organization | 5 | 5 | 5 | 2 | 1 | 3 | 0 | 1 | 22 |
|  | Set the scope of change in the organization | 3 | 4 | 5 | 2 | 1 | 4 | 0 | 1 | 20 |
|  | Identify and communicate the organization's business goals to the staff | 6 | 4 | 4 | 3 | 0 | 3 | 1 | 2 | 23 |

When we looked at the distribution of the responses for supporting process improvements size-wise, we did not find particular size organization supporting process improvements more than the other. Table 5 illustrates, similarly to Table 4, that the tactical level support is more apparent than the strategic, regardless of the size of the organization.

Next, we looked at the responses where standard or model based process assessments had been conducted to measure the process progress (n=19, see Table 2). Table 6 describes how process support was provided in the organizations conducting process assessments. Every support step in the left column also provides the number for the entire sample taking this particular step ("n"). While the strategic support was the least apparent in the overall responses, as indicated in Table 4, the respondents that conducted standard or model based process assessment prepared their organizations for process improvement slightly higher on the tactical levels than on the strategic level. Almost all respondents identify improvement goals and the scope (95%). When we compare the results of Table 6 (those conducting model/standard based process assessment, n=19) with the results of Table 3 (whole sample, n=50), we detect higher percentages for those following a model or a standard (17.4% higher on average across all levels). Almost no difference can be found between the tactical level (+17%) and the strategic level (+17,8%).

**Table 5.** organizational support to process improvement by organizations' size

| | | Micro n=5 | Small n=5 | Medium n=9 | Large n=31 |
|---|---|---|---|---|---|
| *tactical level* | Identify the process improvement goals n=41 | 3 | 5 | 6 | 27 |
| | Identify the process improvement scope n=38 | 4 | 4 | 7 | 23 |
| | Allocate the roles and responsibilities for the improvements n=30 | 3 | 4 | 4 | 19 |
| | Conducting process assessment n=25 | 2 | 1 | 3 | 19 |
| *strategic level* | Get management's support and commitment n=31 | 2 | 2 | 5 | 22 |
| | Decide upon the change strategy for the organization n=22 | 1 | 3 | 2 | 16 |
| | Set the scope of change in the organization n=20 | 2 | 1 | 2 | 15 |
| | Identify and communicate the organization's business goals to the staff n=23 | 1 | 4 | 3 | 15 |

**Table 6.** supporting process improvement in cases where standard or model based assessments were conducted

| | | Conducting model/standard based process assessments n=19 (see Table 2) Count | % |
|---|---|---|---|
| *tactical level* | Identify the process improvement goals n=41 | 18 | 95% |
| | Identify the process improvement scope n=38 | 18 | 95% |
| | Allocate the roles and responsibilities for the improvements in the organization n=30 | 15 | 79% |
| | Get management's support and commitment for the improvement n=31 | 15 | 79% |
| *strategic level* | Decide upon the change strategy for the organization n=22 | 12 | 63% |
| | Set the scope of change in the organization n=20 | 12 | 63% |
| | Identify and communicate the organization's business goals to the staff n=23 | 11 | 58% |

ISO/IEC 33014 argues that the organizations using ISO 9000 are more prepared for and support process improvements better. We looked at the survey data to see whether models and frameworks affect the process improvement support on the strategic and tactical levels. Table 7 illustrates the standards, models and frameworks applied in the respondents' organizations. The table figures indicate how many users of a certain model or framework support the process improvements on the strategic and tactical levels as described in ISO/IEC 33014. The number of all respondents using a model or supporting process improvements is indicated as "n".

We can see that process improvement support on the tactical level is well provided in most cases. From the survey respondents, all CoBIT and PSP/TSP identify process improvement goals; all ISO/IEC 20000 and PSP/TSP users set the scope for process improvements; all ISO/IEC 20000 and Six Sigma users allocate the roles and responsibilities. Most process assessments were conducted in the organizations using ISO/IEC 20000. There is a sharp fall in supporting process improvements on the strategic level aligned to the findings described in Table 4 but the respondents using ISO/IEC 20000 and Six Sigma are providing more strategic support than the other model and framework users. Setting the scope and deciding upon the strategy for organizational change were the most difficult steps on the strategic level, as indicated in Table 4. These steps were taken mostly by the respondents from organizations using Six Sigma.

While the tactical level support for process improvement is provided regardless of the model or framework used in the organization, the organizations using Six Sigma and ISO/IEC 20000 support process improvements on the strategic level more than other model and framework users. Process model users could enhance their organizational support on the strategic level by implementing the management frameworks.

**Table 7.** Models and frameworks & organizational support

| | Our own experience and knowledge | PSP/TSP | ISO/IEC 15504 | CMMI n | Six Sigma | ITIL | CoBIT | ISO/IEC 20000 | ISO 9000 | Lean | Theory of Constraints |
|---|---|---|---|---|---|---|---|---|---|---|---|
| n | 33 | 6 | 15 | 21 | 12 | 24 | 7 | 6 | 17 | 18 | 5 |
| Process improvement goals n=41 | 85% | 100%* | 80% | 95% | 92% | 79% | 100%* | 83% | 94% | 83% | 100% |
| Process improvement scope n=38 | 85% | 100%* | 85% | 90% | 75% | 75% | 86% | 100%* | 82% | 72% | 100% |
| Allocate roles and responsibilities n=30 | 61% | 67% | 73% | 67% | 100%* | 62% | 57% | 100%* | 76% | 83% | 80% |

**Table 7.** (*continued*)

| | Our own experience and knowledge | PSP/TSP | ISO/IEC 15504 | CMMI | Six Sigma | ITIL | CoBIT | ISO/IEC 20000 | ISO 9000 | Lean | Theory of Constraints |
|---|---|---|---|---|---|---|---|---|---|---|---|
| n | 33 | 6 | 15 | 21 | 12 | 24 | 7 | 6 | 17 | 18 | 5 |
| Conduct process assessment n=25 | 48% | 33% | 73% | 71% | 75% | 50% | 71% | 83%* | 71% | 78% | 60% |
| Identify and communicate the organization's business goals n=23 | 45% | 50% | 53% | 48% | 67%* | 50% | 43% | 83%* | 59% | 55% | 60% |
| Decide upon the change strategy for the organization n=22 | 51% | 67% | 60% | 57% | 75%* | 42% | 57% | 50% | 47% | 50% | 60% |
| Management's support and commitment n=31 | 61% | 67% | 80% | 71% | 92%* | 67% | 57% | 100%* | 82% | 78% | 100%* |
| Scope of organizational change n=20 | 42% | 67%* | 47% | 52% | 58%* | 33% | 28% | 50% | 41% | 44% | 60% |

## 4 Conclusions

An international standard ISO/IEC 33014 is currently being developed that describes how organizations can provide organizational support to process improvements. They define eight steps that organizations can take to support improvements on the strategic and tactical levels. As a result of our international survey among the process improvement practitioners, we conclude that the process improvement support is provided in the organizations regardless of their business area and their size. Process assessment revisits the organization's business goals and helps align them to process improvement goals, thus supporting process improvements more on the strategic level. Organizations using process models together with the management frameworks are more likely to support process improvements on both the strategic and tactical levels, increasing their chances to succeed in process improvement initiatives.

Inferential statistics will be done in the near future to see whether the findings can be generalized. But we expect that due to the small sample size, a generalization from the sample to the whole population will be difficult. Nevertheless, the data provides interesting insights, which can be used to design future studies that focus on more concrete aspects.

**Acknowledgments.** The present project is supported by the National Research Fund, Luxembourg and cofounded under the Marie Curie Actions of the European Commission (FP7-COFUND).

# References

1. Arent, J.: Transforming Software Organizations with the Capability Maturity Model. In: Proceedings of the Product Focused Software Process Improvement at PROFES, pp. 103–113 (2000)
2. Lepasaar, M., Varkoi, T., Jaakkola, H.: Models and success factors of process change. In: Product Focused Software Development and Process Improvement, PROFES (2001)
3. Paulk, M.C., Weber, C., Chrissis, M.B.: The Capability Maturity Model for Software. In: Elements of Software Process Assessment and Improvement, pp. 3–15. IEEE Computer Society Press, Los Alamitos (1999)
4. Varkoi, T.: Software Process Improvement Priorities in Small Enterprises. Information Technology, vol. Licentiate of Technology. Tampere University of Technology, Pori (2000)
5. Humphrey, S.W.: Managing the Software Process. Software Engineering Institute (1989)
6. Grover, V.: From Business Reengineering to Business Process Change Management: A Longitudinal Study of Trends and Practices. IEEE Transactions on Engineering Management 46, 36–46 (1999)
7. Zahran, S.: Software Process Improvement - Practical Guidelines for Business Success. Addison-Wesley Professional, Reading (1998)
8. Statz, J.: Measurement for Process Improvement v.1.0. Practical Software and Systems Measurement (2005)
9. Morten Korsaa, M.B., Messnarz, R., Johansen, J., Vohwinkel, D., Nevalainen, R., Schweigert, T.: The SPI manifesto and the ECQA SPI manager certification scheme. Journal of Software Maintenance and Evolution: Research and Practice (2010)
10. ISO/IEC JTC1 SC7: ISO/IEC PDTR 33014.4 - Information technology — Process assessment — Guide for process improvement (2010)
11. Pries-Heje, J., Johansen, J.: Improve IT- A book for improving software projects. Copenhagen (2007)
12. ISO/IEC 15504-2, Information technology - Software process assessment - A reference model for processes and process capability. 15504, vol. 2 (2004)
13. McFeeley, B.: IDEAL: A User's Guide for Software Process Improvement (1996)
14. Cunningham, J.B.: Case study principles for different types of cases. Quality and Quantity 31, 401–423 (1997)
15. Barafort, B., Di Renzo, B., Merlan, O.: Benefits Resulting from the Combined Use of ISO/IEC 15504 with the Information Technology Infrastructure Library (ITIL). In: Oivo, M., Komi-Sirviö, S. (eds.) PROFES 2002. LNCS, vol. 2559, pp. 314–325. Springer, Heidelberg (2002)
16. Jones, C.: The economics of software process improvement. Computer 29, 95–97 (1996)

# Towards a Systemic Maturity Model for Public Software Ecosystems

Angela M. Alves[1,2], Marcelo Pessoa[2], and Clênio F. Salviano[1]

[1] Centro de Tecnologia da Informação Renato Archer, Campinas, Brazil
{angela.alves,clenio.salviano}@cti.gov.br
[2] Universidade São Paulo (USP), São Paulo, Brazil
marcelo.pessoa@usp.br

**Abstract.** Brazilian Public Software (BPS) is an innovative experience in public administration. It combines features of the free software production model with the concept of public goods and is delivered by a portal that links different people and interests. The evolution of BPS as a Public Software Ecosystem (PSE) can be best understood using Complex Thinking Theory (CTT). The papers describes how methodologies based on System Thinking, were used to obtain empirical evidence that the BPS ecosystem evolves in learning cycles and concludes that this could result in a systemic maturity model for BPS, provide a reference for understanding and improving BPS and others PSE. This maturity model has been developed using an analogy with ISO/IEC 15504 (SPICE) references for capability maturity models. The System Thinking however indicated different path for maturity other than the one based on capability. This finding is consistent with the current evolution of SPICE (ISO/IEC 33000 series) towards other path for maturity in addition to capability.

## 1 Introduction

Brazilian Public Software (BPS) is an initiative of the Planning, Budget and Management Ministry (PBMM) of Brazil that introduces a new concept and operational structure to produce software, aimed at improving governmental system efficiency. This initiative began officially in 2006 [1]. At that time, Free Software Production Model (FSPM) adopted by the Federal Government was strongly stimulated by national policies. One of the experiences of code opening, the Cacic infrastructure inventory software [2], gave to Brazilian policy makers the perception that the government and several sectors of the Brazilian society were interested in sharing and improving the software knowledge.

Then, policy makers of PBMM developed the concept of public software and created BPS. BPS is based on FSPM, but it also includes additional duties to the entity interested in making its software available as a public good [3]. This is basically because the government has the legal responsibility to make public goods available (including software) in minimal conditions of use, security and trust. These duties are established by a formal term between PBMM and the entity. This formal term includes: 1) to license software in GPL (General Public License) form; 2) to provide software guidebooks to users and guarantee that software work after installation; 3) to provide a focal point or a team that constitutes a communication interface with the society, driving and solving their needs concerned with the available software; 4) to provide associated services to do the communication with society, like forum, internet site, software version control tools and 5) to manage the collaboration with the virtual community, inducing participating, registering contributions to software, defining quality patterns and launching new versions. To support and to make the concept operational, a virtual ambience portal was created and became operational in 2006. Today there are more than 100.000 people using the portal and 44 solutions available. Around each software solution a community has been formed that interacts with the leaders. Many people participate in more than one community. The solutions came mostly from public entities, but private enterprises began to release software. The growth of software communities in the BPS portal quickly gave rise to demands that in their turn led to new dimensions to be analyzed and incorporated into the BPS model [3]. Dimensions quite different in terms of their nature, such as intellectual property, services commercialization derived from the apprenticeship in the communities, demands on the infrastructure usability, flexibility and interoperability, new policy acts to complete the public good concept even the implantation of the model in other countries in Latin America. In 2009 the researchers from the Information Technology Center Renato Archer (CTI – *Centro de Tecnologia da Informação*) was contacted by PBMM in order to comprehend the complex dynamic of BPS and then elaborate a Reference Model to manage and evolve the ecosystem. This article presents the trajectory followed by researchers in order to solve this research problem. This paper has 7 sessions including this introduction. The session 2 presents the motivation and the context, session 3 - the objectives, methodology and process, session 4 - the path towards a systemic maturity model, session 5 - the model, session 6 - the initial validations and session 7 presents the conclusions.

## 2 Motivation and Context

The most common approach to quality (as for example, ISO/IEC 15504 [17] and CMMI [4]) is based on formally established organizations, with production processes that can be decomposed into smaller units in a linear fashion. So the maturity of an organization is directly proportional to the control of its processes and ability to achieve the goals envisaged. According to the Software Engineering Institute (SEI) a software process can be defined as a set of activities, methods, practices and transformations that people use to develop and maintain software and its associated products (plans and design documents, code, test cases and user manuals). As an organization becomes mature, the software process becomes better defined, allowing it to be implemented more consistently across the organization [4, 17]. This approach is known as Maturity based on Capability.

The word maturity can be used as a more generic term to characterize any approach to identify best practices and learning paths, and organize them as continuous improvement cycles. This work in progress uses the word maturity for a maturity based on systemic thinking instead of capability. The motivation of this work was to develop a maturity model for a complex system like the BPS since the existing capability maturity models were created from the assessment of precisely defined structure, with defined processes, interfaces with the external environment quantified and qualified. For a system such as the BPS, with undefined limits, non-linear relationships and dynamics, the way to build a maturity model had to be different and innovative.

The System Thinking indicated different path for maturity other than the one based on capability. This finding is consistent with the current evolution of SPICE (ISO/IEC 33000 series) [18] towards other path for maturity in addition to capability. The current version of ISO/IEC 15504 establishes process capability as the path for the improvement of a process. The new version (ISO/IEC 33000) confirms process capability as an important path, but recognizes the need for others paths. Therefore ISO/IEC 33000 will established requirement for process measurement framework [18]. Process capability is one process measurement framework. The design, use and validation of the reference model are financed by Brazilian Innovation Agency (FINEP) and are being carried out by CTI as a doctoral research at the University of São Paulo (USP) with completion scheduled in late 2011.

## 3 Objectives, Methodology and Process

The main objective of this article is to present an innovative experience on using free software in public administration as an emergent ecosystem and an attempt to establish a maturity model to that. A complementary objective is to provide an experience that can be used as subsidy for ISO/IEC 33000 requirements for process measurement frameworks. This complementary objective was included during the work.

The methodology used to obtain an appropriate model to the BPS ecosystem was the Collaborative Action Research (AR) [5]. According Thiollent [6], AR is a kind of empirically based social research that is designed and carried out in close association with an action or resolving a collective problem and in which researchers and representative participants of the situation or the problem are involved in a cooperative or participatory way. AR is a research strategy in production engineering which aims to produce knowledge and resolve a practical problem. AR should follow four broad phases: 1) the exploratory phase, 2) the research phase, 3) the action phase and 4) evaluation phase. This work has been conducted since January 2009 following a process based on AR phases. The first phase aimed to define the context and purpose of the problem. The second phase defined the conceptual framework and theoretical research. The third phase has defined and implemented the action plan. The fourth phase included the codification of theoretical and practical results from several cycles performed in the previous phase.

## 4 The Path towards a Systemic Maturity Model

This section describes the main actions and results from the Action Research process. It covers a view on the BPS, the search for a reference model for the BPS, the first attempts towards a maturity model based on learning cycles, the development of a systemic map for the BPS, the movement from the systemic map to a systemic maturity model, and the generalization from the BPS to a Public Software Ecosystem (PSE).

## 4.1 Brazilian Public Software

The BPS model is an alternative use of the FSPM. However, like the FSPM, the public software network tends to the behavior of an emergent network, which means a network with non-centralized growth, without a central coordination, majority of horizontal connections, non-linear results and behaviors [7]. BPS project is inspired by digital ecosystems as broader concept. The term ecosystem is derived from the concept of digital ecosystems, originated in Europe in the late 90's and which defines a conceptual framework to describe the complex interactions between business, technology and knowledge, which is inspired by biological ecosystems [8]. In the context of the BPS, ecosystem concept assumes a distinct, since, although the achievement of business is also included among its objectives, its main focus is on improving public management and public access to IT knowledge, targeted for development sustainable country. Thus, the term ecosystem is used to describe the virtual environment of the BPS [9] consisting of its various actors who interrelate collaboratively and with their own dynamics, not to set limits to the variety of actors or the intensity or forms of these relationships. Thus, we can say that the BPS ecosystem is composed of elements that relate to the infrastructure that supports it, the actors that comprise it and the relations between them and those with this infrastructure. The resulting dynamics is expressed through interaction, sharing and learning is what characterizes this ecosystem. Figure 1 illustrates the range of internal and external relationships to the BPS ecosystem.

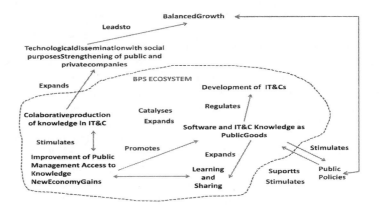

**Fig. 1.** A preliminary view of BPS Ecosystem

## 4.2 Reference Model for Brazilian Public Software

The reference model (RM) to be developed should include the specific nature of the BPS. The BPS is a complex adaptive system and a digital ecosystem of software production and the theoretical basis for the complex adaptive systems and systems thinking were considered adequate to address the model elaboration. The reference model developed includes two views on the ecosystem: top down and bottom up. The model bottom-up consists of five vectors related to Interoperability, Software testing,

Product quality, Quality of service and Quality development. For the top-down view was elaborated systemic maturity model, based on learning cycles of ecosystem BPS. The aim of the maturity model is to support the understanding of the steps taken by the BPS and guide the continuation of this path to achieve their goals. This work does not address the quality model bottom up.

### 4.3 First Attempts towards a Maturity Model: Leaning Cycles

This section aims at analyzing the process of construction and utilization of the maturity model from the viewpoint of a learning process. This analysis is a preparation for a discussion on how the proposed learning cycles presented can lead to a mature model of BPS using an evolution and generalization of the concept of maturity.

In the BPS the maturity levels formulation occurred from the perspective of the four cycles of learning [9, 10]. The perception of these cycles and its conception are based on the CTT view and principles. They are not separated but are inserted in circles of causation, they are present in each community, working group as well as in the ecosystem as a whole, and are inspired on the use of Peter Senge [16] theoretical contributions to learning organizations. So, considering the BPS experience, there are 4 cycles of apprenticeship: I) learning with the structural elements of BPS; II) learning with the components of BPS; III) learning with the relationships of the ecosystem and IV) learning with the patterns of behavior of BPS. Therefore these cycles occur in a sequence, they aren't disconnected and discontinuous. The first cycle continues to occur in the second cycle but in a minor intensity and the third cycle occurs in the first and second cycle, but is not the focus. The figure 2 expresses these considerations.

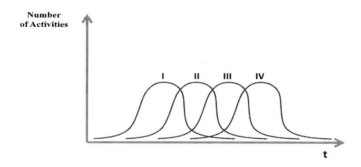

**Fig. 2.** BPS learning cycles

It is important to point out that all these cycles begin from the conceptual, legal and organizational framework of public software. This framework has created and continues to maintain the environment that enables the interactions in the ambiance of BPS. It also acts like an active frontier that, at the same time, promotes interactions with the all stakeholders, absorbing their contributions but maintaining the main objectives of BPS ecosystem. The end of the fourth cycle generates feedbacks to redesign the basic structure of BPS ecosystem and promotes a new beginning of apprenticeship cycles, however in a new level, like a spiral moving. So, each set of cycles contributes to the evolution of the framework of BPS, improves its strategic view, and incorporates new variables, and so on. It is natural to look at these cycles as a process towards increasing maturity that indicate a base to establish a mature model.

### 4.4 Systemic Map for Brazilian Public Software

The choice for CTT approach by the project coordination aimed to integrate different theoretical contributions, some with linear visions, to subjective perceptions and tacit knowledge to describe a quick and dynamic process from which results emerge. So, CTT offers useful cognitive tools to look, understand and to deal with the non-linearity of BPS ecosystem. The use of CTT in the BPS project involved the consideration of some cognitive operators (circularity, feedback mechanisms and dialogic operators) [12, 13]. The use of these operators to build a systemic view of BPS, was based on the approaches described by Senge (1990) [11] and Andrade (2006) [14]. In practical terms, the study of BPS ecosystem drove by CTI produced a systemic map of the BPS. The preliminary results obtained fostered many debates and new insights and learning about the nature of the BPS. The following map agglutinates all data survey and reflections conducted. This map seeks to express a didactic way the complex relationships that occur in the BPS. But the interpretation of this map is too extensive to be treated here. It was built with the participation of key actors who participate in the BPS today: government, community leaders, users, researchers and entrepreneurs. The systemic map was then analyzed from the learning cycles point of view discussed in section 4.3. The map is represented in the Figure 3.

### 4.5 From Systemic Map to a Systemic Maturity Model

From the analysis of the BPS purpose and vision, its critical variables, the systemic map and mental models, we found that the critical variables have some common denominators. These denominators seem to relate to particular learning that some themes provide. Thus, it was observed that a described group of critical variables was related to learning to the infrastructure problems of the BPS and its artifacts. Other variables related to the interfaces of the relationship of BPS. These observations have influenced the formulation of the BPS ecosystem reference model, which reflects the vision and purpose of the BPS, as described in previous sections and from reflections on the construction of the reference model also has advanced the understanding of the maturity model. This process of feedback between the models and the collective construction of the map is extremely similar to systemic process that is observed in virtual communities of BPS, i.e., the process of constructing knowledge about the object (BPS) is similar to his dynamics of growth and evolution.

The analysis of the critical variables of systemic map led to the identification of initially three layers grouping variables with similar characteristics. Each layer represents a learning cycle with a certain aspect of the evolution of BPS. Figure 3 illustrates this exercise[1]. The outer layer (dotted lines) is about to critical variables focused

---

[1] It is not possible due to the size and complexity, maintaining the sharpness of the mental map (Figure 3). The authors decided to expose a partial zoom of the map in order to show part of the variables and relationships. To get the complete picture, contact the authors.

on aspects of infrastructure and artifacts. The inner layer (discontinuous lines), concerning aspects of the BPS relationship with their environment (media, partnerships) and the solid lines layer concerns aspects related to the impacts achieved by the BPS (training, business). So we can say that were initially identified three learning cycles that occur in the BPS. To assess the ecosystem BPS level of maturity, we control the critical variables that led to each level of maturity. From the values assumed by each variable is defined quantitatively and qualitatively the BPS level of maturity.

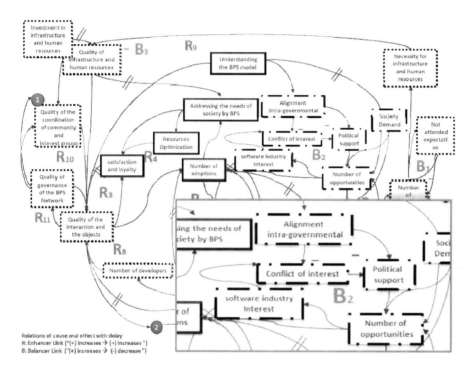

**Fig. 3.** BPS Systemic Map and critical variables

### 4.6 Generalizing from BPS to Public Software Ecosystems

As discussed previously the BPS project is inspired by digital ecosystems as broader concept and we can say that the BPS ecosystem is composed of elements that relate to the infrastructure that supports it, the actors that comprise it and the relations between them and those with this infrastructure. The resulting dynamics is expressed through interaction, sharing and learning is what characterizes this ecosystem. Then we can say that the BPS is a particular case of a public software ecosystem.

## 5 The Proposed Systemic Maturity Model

The maturity model of BPS incorporates normative vision and the levels that must be crossed to reach the desired vision. The maturity model consists of five levels and Figure 4 summarizes each of these levels [10]. The maturity levels are defined from the values, qualitative and quantitative, undertaken by a number of critical variables that were set from the systemic map (detailed in items .4.4 and 4.5). At this first stage were defined 6 variables for level 1, 11 variables for level 2, 7 variables for level 3, 7 variables for the level 4 and 5 variables for level 5. Presented below are the variables that characterize each level of maturity, being, however, this only a first exercise that can be improved over time. The variables are:

a. **Level 1**: conceptual definition, legal framework, computational infrastructure, portal project, support human resources, project leader (champion);
b. **Level 2**: number of accesses, not attended expectation, necessity for infrastructure and human resources, quality of infrastructure and human resources, investment in infrastructure and human resources, quality of the coordination of community and interest groups, quality of governance of the bps network, quality of the interaction and the objects, user satisfaction and loyalty, number of developers, new solutions and interest groups;
c. **Level 3**:dissemination; number of opportunities; society demand; political support; conflict of interest; partnerships; software industry interest;
d. **Level 4**: addressing the needs of society by bps; number of adoptions; business; qualification; number of providers in the virtual public market, resources optimization, understanding the BPS model;
e. **Level 5**: bps replication, component of public policies, budget autonomy, collegiate management, intra-governmental alignment.

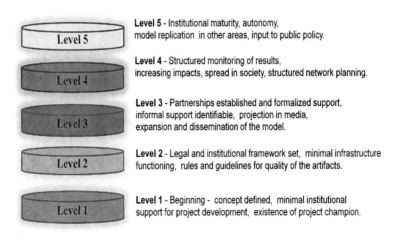

**Fig. 4.** BPS maturity levels

For structured data and surveys conducted within the BPS, we can affirm that the ecosystem is between level 2 and 3. To that assertion more property is needed both to improve the methodology as well as its application in the field. The maturity levels were evaluated using as reference the critical variables that led to each level of maturity. The development process of the reference model was made by collaboratively and shared way. Different actors, from different positions in the ecosystem, participated in the process. A constant during development was the diversity of visions of the actors from the maturity of the same area of the ecosystem. This dichotomy was treated as an emergent process. The dichotomy has been turned into a new dimension that must be aggregated to the model. This dimension is composed of a bottom-up view and a top-down view. The same actor can have both views simultaneously. An actor with bottom-up vision is more sensitive and is more concerned with issues of shorter term or more operational. Actors with top-down view have a level of concern over the long term, more strategic and political.

## 6 Initial Validation

The validation of a maturity model is very important and very difficult activity [20]. Validation also requires long period of time. As this work is developing an innovative maturity model, using an innovative perspective for maturity models, in this case, system thinking, a full validation is not feasible yet. However, in order to gain confidence in the quality of the work, initial and partial validations were identified and used.

**A first initial and partial validation is using an analogy.** With the success of SPI and the most relevant models used as reference for SPI (CMM, CMMI and ISO/IEC 15504 models), mostnew models adapt one of these models. However another approach could be the identification of the process used to produce a model, as for example, the original CMM model, and the usage of an adapted process to generate another model. Such approach is more coherent with process improvement. In this sense a fundamental point in the process used to develop the original CMM model was identified and related with the development of this systemic maturity model. In the 1980´s, Watts Humphrey (SEI) developed a framework to help software development organizations improve their software process. That framework was based on capable software groups that develop software for the USA federal government. "In deciding how to evaluate software organizations, we first listed the characteristics of capable software groups. This produced a set of about 100 questions. Since we expected these evaluations to be performed by Air Force personnel and not by software experts, the questions dealt with specific activities that the assessors could readily understand and verify. Once we had the questions we faced another problem: how could people who were not software experts use the questionnaire to evaluate and rank organizations? …. It was early October in 1986 and I had lots of time on my hands, so I decided to work on this problem. It occurred to me to list these 100 questions against Crosby's maturity framework. [In 1979, Philip B. Crosby proposed a Quality Management Maturity Grid as five mature levels in terms of how mature an organization´s processes are, and how well they are embedded in their culture, with respect to service or product quality management]. To my surprise, they fit very well.

While I had to redefine the maturity levels, it was not hard to rank the questions in maturity order" [19, p.47]. Similar action with similar results was performed in the development of this systemic maturity model. The learning cycles proposed as a first attempt towards a maturity model acted as the Crosby´s Quality Management Maturity Grid. The BPS´s systemic map and its critical variables act as the "SEI´s 100 questions". The movement from the systemic map to a systemic maturity model acts as the listing of "these 100 questions against Crosby's maturity framework". Also in this case, while we had to adjust the maturity levels, it was not hard to assign each cycle and variable into the maturity levels.

**A second initial and partial validation is the usage of Action research.** The work of preparing the reference model of the BPS had practical and theoretical questions. Theoretical issues were related to weaknesses that difficult the treatment and understanding of digital ecosystem for software production such as BPS. Because of these characteristics the researchers chose to use the AR methodology that includes the two dimensions of the problem (practical and theoretical). The exploratory phase of research to diagnosed the problems in the ecosystem, in research phase were selected theoretical frameworks to be used and drew up an research action plan of, in the action phase the plan was implemented in seven cycles. In the assessment phase of the research, it was found that the actions taken by the AR met the expectations of actors in the BPS. The knowledge generated by the process was incorporated into the ecosystem and its actors. A model was generated and is used as reference by the ecosystem. From the viewpoint of the research was generated sufficient knowledge for the development of a maturity model for public software ecosystems. This maturity model is innovative because it moves from the maturity model for organizations with contours, objects and functions clearly defined to other type of arrangement, also productive, that are the public software ecosystems. The research has generated a second innovation that is the incorporation of multiple views of the maturity levels, which depends on the position occupied by the actor, within the same ecosystem.

**A third initial and partial validation is a cooperative dimension of a software development model.** In another related project, a process capability/cooperative reference model for software development in BPS context has been developed. From an analyses of the current communities for the forty-four software systems in BPS, and from a reflection on previous experience, we concluded that there are many approaches to coordinate each BPS community and they vary in a range from emphasis in "command and control" to emphasis in "cooperation and connection". Therefore the reference model should provide best practices and improvement paths that allow each community to choose a "point" in this "range" and guides the improvement based in this choice. As for an emphasis in "command and control" there are already the capability levels established by ISO/IEC 15504 (SPICE), they were included in the model as the capability dimension. The path developed for this systemic maturity model has been adapted and the result is included in this model as a cooperative dimension. The development of this cooperative dimension based on the systemic maturity model paths provides some confidence about the quality of the systemic maturity model. Therefore this development is another initial and partial validation for the systemic maturity model.

**A fourth initial and partial validation is a solution of mismatch expectations for a project.** After presenting the work Brazilian Public Software: beyond Sharing[15] at The International ACM Conference on Management of Emergent Digital EcoSystems in 2010/October the BPS reference model development team was invited by FP7: Seventh Framework Programme to submit a proposal for the call FP7-ICT-2011-EU-Brazil [16]. The project should have, as its main objective, to replicate the Brazilian experience, the public digital ecosystem of software in European cities. During the process of drafting and discussing the project with European partners the team identified a mismatch between the team strategy and the partners´ strategy. As we were using the systemic maturity model to understanding and indicating the focus of the project, we decide to try to understand the partner position in terms of the systemic maturity model. The result was a clear understanding where the mismatches were. The demands placed by the partners with respect to the ecosystem were at a level of maturity well beyond level that the ecosystem is today. The current ecosystem is at level 2. We were focus the project in improve it to the level 3, because this was a challenge feasible objective for the budget and time frame of the project. The partners demands were to improve the ecosystem to level 5, what is not feasible. Therefore the systemic maturity model helped the construct of a better understanding.

# 7 Conclusions

The maturity models traditionally aim to improve the quality of a linear and well defined process. In the case of a virtual network, whose design is closer to a cloud, with no defined boundaries, roles that change, mutant settings, the conventional view of maturity model is not appropriate. The same goes for the BPS. To think a maturity model for the BPS was necessary to use concepts from the CTT. This approach allowed us to build an overview of the BPS and how the variables interrelate and interact. The analysis of the systemic map and critical variables identified the BPS learning cycles that defines and sets its internal dynamics. The maturity levels were defined based on the learning cycles. The reference model of the BPS ecosystem is a model built to drive the evolution of the ecosystem as well as to allow a new user construct an integrated view of the various elements that compose it. The constructed model ecosystems can be generalized to public software ecosystems and incorporates an innovation: the model has incorporated the dimension of multiple views of the level of maturity. The model allows evaluating the maturity of ecosystems with an emphasis on cooperation and connection, unlike the maturity model known and used by organizations which are based on command and control. The positive results from the four initial and partial validations performed so far gave enough confidence that this systemic maturity model is going in a promising direction. For SPICE community the process performed and the current results could be used for the emergent ISO/IEC 33000 requirement for process measurement framework. In the other hand, this systemic maturity model should use this ISO/IEC 33000 requirements, when they be published, as a reference for this model future consolidation.

# References

[1] Freitas, C.S., Meffe, C.: FLOSS in an Open World: best practices from Brazil. In: 2020 FLOSS Roadmap, 01st edn. vol. 01, pp. 69–73. Creative Commons, Paris (2008)
[2] Sistema de Inventário CACIC, http://www.softwarepublico.gov.br/dotlrn/clubs/cacic (last access MAI/2010)
[3] Meffe, Corinto: O avanço do Software Público Brasileiro Experiência Brasileira Linux Magazine 49, 28-32 (December 2008)
[4] CMMI Product Team, CMMI® for Development, Version 1.3, Improving processes for developing better products and services. TECHNICAL REPORT, CMU/SEI-2010-TR-033, ESC-TR-2010-033, Software Engineering Process Management Program (November 2010)
[5] Coughlan, P., Coghlan, D.: Action research for operations management. International Journal of Operations& Production Management 22(2), 220–240 (2002)
[6] Thiollent, M.: Metodologia da pesquisa-ação, 14ª edn. São Paulo, Cortez (2005)
[7] Castells, M.: The Rise of the Network Society, 2nd edn. Blackwell Publishing, U.S (2000)
[8] Briscoe, G., De Wilde, P.: Digital Ecosystems:Self-Organisation of Evolving Agent Populations. In: Proceedings of the International Conference on Management of Emergent Digital Ecosystems International Conference on Management of Emergent Digital Eco Systems. ACM, Nova Iorque (2009)
[9] Alves, A.M., et al.: Learning path to an emergent ecosystem: the Brazilian public software experience. In: Proceedings of the International Conference on Management of Emergent Digital EcoSystems International Conference on Management of Emergent Digital EcoSystems, Lion, França. ACM, Nova Iorque (2009)
[10] Alves, A.M., Pessôa, et al.: Systemic Maturity Model and Brazilian Public Software. IEEE/PIC-2010 - International Conference on Progress in Informatics and Computing – Shanghai, China, de dezembro 10-13 (2010)
[11] Senge, P.M.: The Fifth Discipline: The Art & Practice of Learning Organization. Doubleday, New York (1990)
[12] Morin, E.: Introduction à la pensée complexe. Est Editeurs, Paris (1990)
[13] Maturana, H.R., Varela, F.J.: Autopoiesis: The organization of the living. In: Maturana, H.R., Varela, F.J. (eds.) Autopoiesis and cognition. Reidel, Boston (1973)
[14] Andrade, A., Seleme, A., Rodrigues, L., Souto, R.: Pensamento Sistêmico – Caderno de Campo. Bookman, Porto Alegre (2006)
[15] Alves, A.M., Pessoa, M.: Brazilian Public Software: beyond Sharing. In: ACM/MEDES 2010: The International ACM Conference on Management of Emergent Digital EcoSystems –Bangkok, Tailândia, outubro 26– 29, pp. 73–80. ACM Press, New York (2010)
[16] cordis.europa.eu/fp7/dc/index.cfm?fuseaction=usersite.FP7DetailsCallPage&call_id =377
[17] ISO/IEC 15504, composed of seven parts (15504-1 to 15504-7) parts. under the general title Information technology — Process assessment(2004-2008)
[18] Dorling, A.: Next Generation 15504 - the 33001 series of Standards – UPDATE, August 24 (2009), http://www.spiceusergroup.org
[19] Humphrey, W.S.: Three Process Perspectives: Organizations, Teams, and People. In: Annals of Software Engineering, vol. 14, pp. 39–72. Kluwer Academic Publishers, Dordrecht (2002)
[20] von Wangenheim, C.G., et al.: Creating Software Process Capability/Maturity Models. IEEE Software 27(4), 92–94 (2010), doi:10.1109/MS.2010.96

# Linking Software Life Cycle Activities with Product Strategy and Economics: Extending ISO/IEC 12207 with Product Management Best Practices

Fritz Stallinger[1], Robert Neumann[1], Robert Schossleitner[2], and Rene Zeilinger[1]

[1] Software Competence Center Hagenberg, Softwarepark 21, 4232 Hagenberg, Austria
{fritz.stallinger,robert.neumann,rene.zeilinger}@scch.at
[2] STIWA Automation GmbH, Salzburger Straße 52, 4800 Attnang-Puchheim, Austria
robert.schossleitner@stiwa.com

**Abstract.** Traditional software engineering reaches its limits when facing increased complexity and variability of software products. Approaches like software product line engineering are considered promising to successfully tackle such challenges. However, to fully exploit its potential, any product- and reuse-focused development approach poses the need to closely link core software engineering activities with strategic and economic product aspects. Product management is generally expected to bridge this gap from business and product related goals to software life cycle activities, but often fails to deliver the promised outcomes. Moreover, software process improvement approaches generally lack the provision of explicit or detailed product management activities. – This paper presents the results of extracting key product management practices from selected software product line and product management frameworks. The obtained results are compared against ISO/IEC 12207 in order to identify directions for defining a standard conformant reference model for process improvement in product-oriented software development contexts.

**Keywords:** software product management, software product line, software process, process reference model, ISO/IEC 12207, software process improvement.

## 1 Introduction

Traditional, typically project-oriented software development organizations are increasingly confronted with the need to develop or enhance their software as a product.

The implied transition from a project-oriented to a product-oriented development approach typically requires a change of view in the whole organization, poses the need for integration of a series of further stakeholders with potentially diverging views and perceptions into the development and maintenance process, and generally increases the importance of business and market but also technological considerations. Specific challenges include amongst others:

- the coverage of the needs and expectations of many different existing and potentially new customers with the developed products and services,
- the alignment of the products and services to specific markets or market segments,
- the handling of increased functionality, variability, and complexity of products,

- the selection of an appropriate product architecture,
- the development of product variants by reusing existing solutions,
- the enhancement of products in alignment with a product portfolio, and
- the coordination of interdependent and interacting software products or of software as part of other products, e.g. mechatronic systems.

An essential goal of such a transition is to harvest the benefits of pre-defined and – ideally – pre-developed products or product components while still satisfying the individual customers' needs.

A promising approach to realize such benefits is software product line engineering (SPLE) (cf. [1], [2], [3]) which generally enables software developing organizations to gain significant improvements with respect to development and enhancement costs, time-to-market, and product quality.

However, to fully exploit the potential of product- and reuse-focused development approaches, core software engineering activities, e.g. requirements engineering, architecture engineering, or quality assurance, have to be closely linked and aligned with strategic and economic product aspects that are typically derived from or related to the overall business goals of the organization.

## 1.1 Products and Product Management

The term 'product' in our context and in the context of SPLE in general refers to applications denoting both software and software-intensive systems; products may also be services or solutions offered to the customer [4].

Establishing product management is generally considered a key element in the transition towards product-oriented software engineering and in particular with respect to bridging the gap from business and product related goals to software life cycle activities.

Product management is well established in other domains – mainly consumer products, mass products, etc. It is commonly defined as the 'planning, organising, executing, and controlling of all tasks, which aim at a successful conception, production, and marketing of the products offered by a company' [2].

In the context of SPLE, product management 'aims to define the products that will constitute the product line as a whole' [3]. According to these definitions, product management aims at identifying the major commonalities and variabilities among the products in the product line and realizing product portfolio planning accompanied by major economic analysis of the products.

## 1.2 Goals and Approach

The long-term vision of the work presented here is to support the transition of software development organizations towards product- or product line-related development approaches by fostering the establishment of product management as key intermediary between business goals and core software engineering activities.

The overall goal is to provide a best practice process reference model for successful product-oriented software engineering, which is based on a common language for communication among stakeholders and which integrates the topics relevant for organizations aiming to provide solutions for their customers based on software products (including business-related topics like marketing, sales, finance, etc. where appropriate).

The approach chosen to achieve this goal is to develop a process model conformant to the requirements of ISO/IEC 15504 [5] for process reference models that fosters the establishment of product management within software engineering organizations on a process level. Ideally, this should be achieved by integrating the respective product management practices into the existing process reference model of ISO/IEC 12207 [6]. The initial steps in this approach are presented in this paper and comprise:

- the identification of the significant product management activities and outcomes through analyses of existing product-oriented models and frameworks
- the comparison of these activities with ISO/IEC 12207 and identification of covered/not covered outcomes
- the identification of possibilities for integration of missing outcomes into the reference model of ISO/IEC 12207
- the establishment of traceability of the additional outcomes and practices to the respective source models and frameworks.

The remainder of the paper is structured as follows: section 2 briefly presents the characteristics of selected frameworks for software product management, software product line engineering, or product-oriented software development in general; section 3 presents the results of extracting key product management activities from these frameworks; section 4 compares the obtained results against ISO/IEC 12207 in order to judge their coverage by the international standard; finally, section 5, summarizes the obtained results and provides analyses and conclusions towards defining a process reference model for process improvement in product-oriented software development contexts through extension of ISO/IEC 12207.

## 2 Analyzed Frameworks

Four frameworks were selected for analysis.

First, the *reference framework for software product management* developed by van de Weerd et al. [7] is based on literature research and field interviews with experienced product management practitioners from software product companies in the Netherlands. The framework describes the product management domain by means of four process areas and sub-functions, and provides their relations with internal and external stakeholders as well as information flows within the domain. The framework was validated in a case study.

The *Framework for Software Product Line Practice* [8] provided by the Software Engineering Institute (SEI) captures the latest information on successful software product line practices in technical and organizational areas. The provided information is constantly updated based on studies of and direct collaboration with organizations that have built product lines or are involved with software product lines otherwise, and with leading practitioners.

Pohl et al. provide a *software product line engineering framework* [2] which captures the concepts of traditional product line engineering and differentiates between the two processes domain engineering and application engineering. A particular characteristic of this framework is the incorporation of product management as a key sub-process of the domain engineering process.

The fourth framework selected is the *Microsoft Solutions Framework* [9] which aims to support organizations in successfully delivering information technology solutions and technology projects. Within this framework, the *MSF Team Model* defines roles and responsibilities of teams and their members and covers interdependent multi-disciplinary roles within such information technology projects, of which one is product management.

## 3 Identification of Key Product Management Activities

The frameworks identified in section 2 above address product management and SPLE in different ways, e.g. the product management framework of van de Weerd et al. [7] is described in terms of process areas and sub functions of the product management domain, whereas in [8] various practice areas are provided, of which each represents 'a body of work or a collection of activities that an organization must master to successfully carry out the essential work of a product line'. From each framework the topics explicitly associated with product management and those SPLE-related topics considered relevant for product management were used for analysis.

From the resulting set of product management related topics a series of key product management activities was compiled, each covering similar or strongly connected topics from the individual frameworks. Table 1 shows an overview of these key activities and the source frameworks which mainly provide the respective topics.

**Product Portfolio Management** covers decision making about the set of existing and in-development products offered by an organization, including identification, evaluation, selection, and prioritization of products, as well as decision on the products' life cycles [2], [7]. It is a strategic function that aims to establish and maintain a balanced and value maximized product portfolio that supports the business strategy and makes optimal use of the organization's resources. The product portfolio typically contains product types or classes of products instead of all the individual products [2].

**Product Life Cycle Management** is 'a comprehensive approach for product-related information and knowledge management within an enterprise, including planning and controlling of processes that are required for managing data, documents and enterprise resources throughout the entire product life cycle' [7]. The product life cycle describes an idealized progression of a product through different stages. One commonly applied life cycle model is for example based on the sales and profits of a product and is comprised of the life cycle stages introduction, growth, maturity, saturation, and degeneration [2].

**Table 1.** Key product management activities and source frameworks

| Key Product Management Activities | Source Frameworks | | | |
|---|---|---|---|---|
| | van de Weerd et al. [7] | SEI [8] | Pohl et al. [2] | MSF [9] |
| Product Portfolio Management | x | | x | |
| Product Life Cycle Management | x | | | |
| Product Roadmapping | x | x | x | |
| Requirements Engineering | x | x | | |
| Release Planning | x | | | x |
| Market Monitoring | x | x | x | x |
| Domain and Product Line Scoping | x | x | x | |
| Asset Identification | x | x | | |
| Product Planning | x | x | x | x |
| Product Controlling | | x | x | |
| Customer Interface Management | x | x | x | |
| Funding | | x | | |
| Product Innovation | | | x | |
| Cross-functional Communication | | | | x |

**Product Roadmapping** deals with long-term plans and expectations and outlines the products in the portfolio as far as they are foreseeable. The product roadmap determines the major common and variable product features and a schedule with their planned release dates [2]. According to [10], in the software business the roadmap typically addresses a strategic timeframe of up to five years and shows, among features and schedules, also important dependencies of the product to other products or platform technologies.

**Requirements Engineering** comprises the elicitation, analysis, specification, verification, and management of user and product requirements in a systematic and repeatable way and aims to ensure their completeness, consistency, and relevance [8]. Dependencies between requirements and traceability to customers, products, features, or product components are also of high importance.

**Release Planning** covers the definition of product releases by prioritizing and selecting the product requirements to be implemented in each specific release [7]. Each product release has a particular purpose and is made available to specific stakeholders, typically customers but also for example test and quality assurance teams of the organization.

**Market Monitoring** is the observation and analysis of the external factors that determine the success of a product in the marketplace, e.g. customers or customer groups, current and potential competitors, trends of prices, technologies, and buying and usage patterns, and barriers to market entry or exit (e.g. legal restrictions, investment effort) [4], [8].

**Domain and Product Line Scoping** determines the relevant entities within the domain, that products will interact with, and the domain's boundaries. It establishes product commonalities and sets limits to their variability. Domain scoping requires a profound understanding of the domain, i.e. experience and knowledge of the concepts, terminologies, problems, and solutions common to this domain, and ensures that this information is captured and appropriately represented and communicated to stakeholders [8]. In a SPLE-context, this includes the identification of the product line, the products it contains and their major features, commonalities, and variabilities [2].

**Asset Identification** aims to identify and define particular assets and components that cover the commonalities of and are shared by multiple products and thus are developed for reuse [7], [8].

**Product Planning** covers both strategic and technical product planning. It outlines and determines the product-related goals, strategies to achieve these goals, intermediate objectives, activities to be performed to achieve the objectives, and the allocation of resources to these activities [8]. It includes the definition of a business case for the product and identification and evaluation of make-or-buy opportunities for product components [8] as well as the selection of product ideas for realization and definition of the major features of the envisioned product [2].

**Product Controlling** is concerned with monitoring and guiding product related effort to ensure successful achievement of the product's as well as the organization's goals and objectives. It involves defining and refining goals, identifying respective success criteria and indicators, defining appropriate measures, and developing plans to operationalize and verify these measures [8].

**Customer Interface Management** comprises the understanding and management of commitments between an organization's producers and its customers [8], referred to as partnering and contracting in [7]. This involves for example identifying the customer's representatives, the information to be communicated and delivered to customers (e.g. product offerings, variants, costs, schedules, quality, or benefits), policies and procedures that apply to customer interaction, as well as ensuring that people with customer responsibilities in the organization are trained properly [8].

**Funding** covers the activities to plan and establish adequate financing of software development efforts undertaken in the organization, e.g. development and update of reusable assets and components, development of new products, performing analyses, or establishing and modernizing infrastructure. It includes the identification of appropriate funding sources, the definition of funding requirements and models, and it must be sufficient with respect to the desired quality of results [8].

**Product Innovation** aims at the extension of the product portfolio with new or enhanced products that satisfy customer needs. Various strategies (e.g. innovation leader, product imitation, active or passive search for ideas) and sources for idea generation can be utilized [2].

**Cross-functional Communication** describes the activities of product management as an intermediary between various stakeholders or business functions [9], e.g. customers, development and project teams, management, sales, marketing, research and development, or customer support.

## 4 Key Product Management Activities vs. ISO/IEC 12207

The international standard ISO/IEC 12207:2008 Systems and software engineering – Software life cycle processes provides 'a comprehensive set of life cycle processes, activities and tasks for software that is part of a larger system, and for stand alone software products and services' [6]. The standard regards the ubiquity of software within systems engineering and fosters the view of software design as an integral part of system design. Fig. 1 provides an overview of the respective ISO/IEC 12207 life cycle process groups and processes.

ISO/IEC 12207:2008 is an initial result of an ongoing effort to harmonize, align, and integrate system and software life cycle processes. It has a strong relationship to the international standard ISO/IEC 15288:2008 Systems and software engineering – System life cycle processes [11], and can be used without or in conjunction with ISO/IEC 15288. Further, ISO/IEC 12207 provides a process references model that is conformant with the requirements of ISO/IEC 15504.

### 4.1 Initial Observations

In ISO/IEC 12207 the project provides 'the context for describing processes concerned with planning, assessment and control' [6]. The key product management activities on the other hand are described independently of projects in the context of products or product groups as part of an organization's portfolio.

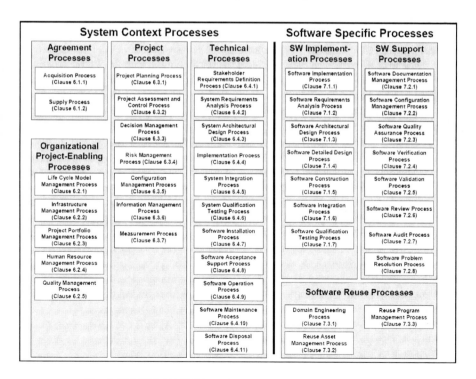

Fig. 1. ISO/IEC 12207 life cycle process groups and processes [6]

ISO/IEC 12207 utilizes a very general product definition – that is a product is 'the result of a process'. A software product is defined as a 'set of computer programs, procedures, and possibly associated documentation and data' [6]. In contrast, from a product management perspective a product can also be a service or solution (cf. 1.1).

### 4.2 Comparison Approach and Results

The comparison of the key product management activities with ISO/IEC 12207 was performed in a one-directional way, i.e. for each activity it was identified to what extent it is covered by processes of ISO/IEC 12207. For each activity the content of its description and detailed topics from the analyzed frameworks were compared against the description of ISO/IEC 12207 processes and process outcomes. The detailed activity and task descriptions of these processes were partly utilized to gain a better understanding of the processes and their outcomes.

For each key product management activity its coverage by ISO/IEC 12207 processes was evaluated using an NPLF-scale (none, partially, largely, fully covered). An overview of the results is provided in Table 2.

ISO/IEC 12207 provides a *Project Portfolio Management Process* which basically addresses the purpose and content of the *Product Portfolio Management* activity, but entirely focuses on projects. Although in ISO/IEC 12207 projects are tightly entangled with products, since a project's purpose is to 'create a product or service in accordance with specified resources and requirements' [6], *Product Portfolio Management* requires a project-independent view, which is only provided in very few parts by the *Project Portfolio Management Process*.

*Product Life Cycle Management* is covered by the *Life Cycle Model Management Process*, which addresses life cycle policies, processes, models, and procedures in general and almost project-independent, but again without a specific product focus.

*Product Roadmapping* is not addressed in any of the ISO/IEC 12207 processes. ISO/IEC 12207 states that the provided processes, activities, and tasks are to be 'applied during the acquisition of a software product or service and during the supply, development, operation, maintenance and disposal of software products' [6], and therefore, product-focused activities, independent of or crossing multiple projects, are not in the scope of the standard.

*Requirements Engineering* is largely covered by the *Stakeholder Requirements Definition Process*, *System Requirements Analysis Process*, and the *Software Requirements Analysis Process*. Traceability of requirements to the various software development artifacts is additionally addressed in the respective software implementation processes (e.g. *Software Detailed Design Process* establishes traceability between detailed design and requirements). From a product management perspective, requirements can be associated with multiple customers and products, variants of products, and assets that are reused in various products. The management of these diverse interrelations and dependencies and maintaining requirements consistency and traceability is not addressed by ISO/IEC 12207 processes.

*Release Planning* is not explicitly addressed in ISO/IEC 12207. Controlling, management, and delivery of releases of software items are covered by the *Software Configuration Management Process* and the existence of a release strategy is implied by the *Software Maintenance Process*. But the concrete planning and definition of product releases in terms of requirements prioritization and selection is not covered.

**Table 2.** Coverage of key product management activities by ISO/IEC 12207 processes

| Key Product Management Activity | NPLF | Evaluation of and ISO/IEC 12207 Processes Mainly Covering the Key Product Management Activity |
|---|---|---|
| Product Portfolio Management | P | Project Portfolio Management Process |
| Product Life Cycle Management | F | Life Cycle Model Management Process |
| Product Roadmapping | N | - |
| Requirements Engineering | L | Stakeholder Requirements Definition Process, System Requirements Analysis Process, Software Requirements Analysis Process |
| Release Planning | N | - |
| Market Monitoring | N | - |
| Domain and Product Line Scoping | L | Domain Engineering Process |
| Asset Identification | F | Domain Engineering Process |
| Product Planning | N | - |
| Product Controlling | P | Measurement Process |
| Customer Interface Management | P | Supply Process |
| Funding | N | - |
| Product Innovation | N | - |
| Cross-functional Communication | P | Stakeholder Requirements Definition Process, Supply Process |

*Market Monitoring* is not covered in ISO/IEC 12207 due to its scope. On a very generic level the *Measurement Process* partly provides appropriate purpose and outcome statements, e.g. to 'collect, analyze, and report data [...] to support effective management' [6], which would also apply for the measurement and analysis of market related data. Currently the *Measurement Process* identifies the information needs of the other ISO/IEC 12207 technical and management processes, of which none requires market related information, and focuses on product, process, and quality data rather than business context information.

*Domain and Product Line Scoping* is largely covered by the *Domain Engineering Process*, whose outcomes include identifying domain boundaries, relationships to other domains, the essential features, concepts, and functions in the domain, and the family of systems within the domain and their commonalities and variabilities.

*Asset Identification* is addressed by the *Domain Engineering Process* in terms of assets that are common to a domain and designed for reuse. Additionally, ISO/IEC 12207 provides a separate *Reuse Asset Management Process* for the management of the identified assets across their entire life cycle.

Project-independent, long-term, strategic, and technical *Product Planning* is not in the scope of ISO/IEC 12207. The *Project Portfolio Management Process* and *Project Planning Process* address similar strategic and planning aspects but on a project level that again is not suitable for product management.

*Product Controlling* basically is not addressed in ISO/IEC 12207, although controlling of projects and problem resolution and changes related to assets and other software items is covered in respective processes (e.g. *Project Assessment and Control Process, Software Problem Resolution Process, Reuse Asset Management Process, Software Configuration Management Process*). The *Measurement Process* in principle covers the respective measurement activities required within *Product Controlling*, but there is no process within ISO/IEC 12207 which requires such project-independent product information (cf. *Market Monitoring* above).

*Customer Interface Management* is partly addressed by the *Supply Process*, which covers the identification of potential customers and the agreement and contracting activities for a concrete product delivery. It does not cover project-independent establishment and management of long-term focused customer relationships.

*Funding* of product-related software development efforts is not covered in ISO/IEC 12207. The *Project Portfolio Management Process* 'commits the investment of adequate organization funding and resources' [6] and thus implies the existence of funding, but again employs a project perspective.

*Product Innovation* in terms of extending the product portfolio with new or enhanced products is not in the scope of ISO/IEC 12207. Various processes handle change requests that are provided by customers, asset users, or other sources and can affect specific products, assets, or other software items. A systematic approach to managing and utilizing these sources, and generating, evaluating, and realizing ideas for new or enhanced products, etc. is not addressed.

*Cross-functional Communication* is addressed in those parts that deal with the communication between teams or roles within development (e.g. architect, domain engineer, asset manager, or implementer) or the communication with the customer and other stakeholders (e.g. *Supply Process, Stakeholder Requirements Definition Process*). Communication with organizational units that are not involved in product development but nevertheless deal with the organization's products, like marketing and sales, finance, or research and development are not covered.

## 5 Summary and Conclusions

The paper presented key product management activities that were extracted from selected software product management and product line frameworks. The comparison of the resulting activities with ISO/IEC 12207 allowed a judgment of the coverage of product management activities by this international standard.

The comparison showed, that life cycle and engineering related activities (*Product Life Cycle Management, Domain and Product Line Scoping, Requirements Engineering, Asset Identification,*) are well covered, which is an expected and not surprising result, and that activities more specific for product management, which are concerned with project-independent, product-specific, or cross-product topics, are not in the scope of ISO/IEC 12207 and therefore just partly or not at all covered by the respective life cycle processes.

Product management activities inherently are cross-functions over the life cycle of products, assets, etc. ISO/IEC 12207 on the other hand, employs a life cycle perspective and aligns the processes not across but along the life cycle of software products. Therefore, these processes address product management activities as a whole or in parts only at the points where they impact the software life cycle. For example, the interface to a customer in ISO/IEC 12207 is addressed by the *Supply Process* for agreement and contracting purposes and later in the life cycle of the software product by the *Stakeholder Requirements Definition Process*. The management of the customer interface outside this context, which product management is additionally concerned with, is not in the scope of ISO/IEC 12207 and therefore not covered.

Further, processes, activities, and tasks within ISO/IEC 12207 imply the existence of particular artifacts (e.g. organizational policies, release strategy, adequate funding, or change requests) which are brought in from outside the life cycle. Product management activities partly aim to provide such artifacts, e.g. developing product strategies and development plans, or generating ideas for product enhancements and thus internally triggering or feeding into specific life cycle activities.

In order to define a standard conformant process reference model for process improvement in product-oriented software development contexts, the incorporation of the key product management activities is intended to be performed at outcome level. Conceptually, for each product management activity outcomes can be defined, which represent the observable result of successful achievement of the activity. These outcomes can be adaptations or specializations of existing outcomes of or additional outcomes for particular ISO/IEC 12207 processes. On the other hand, a product management activity outcome may not be suited to be assigned to an existing life cycle process. In this case, these outcomes provide the basis for additional product management specific processes. – In the following, a brief description of directions on how this incorporation could be established is provided.

The activities *Requirements Engineering, Domain and Product Line Scoping, Asset Identification*, and *Product Life Cycle Management* are well covered by the life cycle processes, and – from the current viewpoint – require no specific integration with existing ISO/IEC 12207 processes.

Outcomes of the *Product Roadmapping* activity can be incorporated to processes involved with long-term or strategic planning aspects (e.g. *Project Portfolio Management Process*, or *Project Planning Process*) and to processes that address product features, commonalities, and variabilities on a high level, e.g. *Domain Engineering Process*.

*Release Planning* outcomes can be linked to life cycle processes which address planning aspects (e.g. *Project Portfolio Management Process, Project Planning Process*) or deal with software requirements (e.g. *Software Requirements Analysis Process, Software Maintenance Process*).

*Product Planning* outcomes can be linked to e.g. the *Supply Process, Domain Engineering Process*, or *Project Portfolio Management Process* with regard to product features, core assets, and long-term planning, to the *Project Planning Process* for development planning, or the *Acquisition Process* for make-or-buy evaluations. Strategy-related outcomes of *Product Planning* that address for example the business case or the product goals and strategy, are not suited for any of the existing life cycle processes.

Similarly, some outcomes of the *Customer Interface Management* activity are suited for the *Supply Process, Software Acceptance and Support Process, Software Operation Process*, and *Software Maintenance Process* since they interface with customers on an operational level. Other outcomes related to the long-term management of customer relationships, communication strategies, etc. need to be linked to an additional process.

The remaining product management activities (*Product Portfolio Management, Market Monitoring, Product Innovation, Cross-functional Communication*, and *Funding*) require additional processes to incorporate their outcomes. With the exception of *Funding*, which is suited for the *Organizational Project-Enabling Processes* group, all of these new processes potentially comprise an additional product-management focused process group.

**Acknowledgements.** The work presented in this paper is supported within the *COMET*-Programme of the *Austrian Research Promotion Agency (FFG - Österreichische Forschungsförderungsgesellschaft)* within the Project *Hephaistos* (Integrated Product Engineering, 2008-ongoing).

# References

1. Clements, P., Northrop, L.: Software Product Lines: Practices and Patterns. In: The SEI Series in Software Engineering. Addison-Wesley, Reading (2002)
2. Pohl, K., Böckle, G., van der Linden, F.: Software Product Line Engineering – Foundations, Principles, and Techniques. Springer, Berlin (2005)
3. van der Linden, F., Schmid, K., Rommes, E.: Software Product Lines in Action. Springer, Berlin (2007)
4. Sabisch, H.: Produkte und Produktgestaltung. In: Kern, W., Schröder, H.-H., Weber, J. (eds.) Handwörterbuch der Produktionswirtschaft, pp. 1439–1450. Schäffer-Poeschel, Stuttgart (1996)
5. ISO/IEC 15504-2. Information Technology – Process Assessment. International Standards Organization (2003)
6. ISO/IEC 12207:2008. Systems and software engineering — Software life cycle processes. International Standards Organization (2008)
7. van de Weerd, I., Brinkkemper, S., Nieuwenhuis, R., Versendaal, J., Bijlsma, L.: Towards a Reference Framework for Software Product Management. In: 14th IEEE International Requirements Engineering Conference, pp. 319–322. IEEE Computer Society, Minneapolis (2006)
8. A Framework for Software Product Line Practice, Version 5.0, http://www.sei.cmu.edu/productlines/frame_report/index.html
9. Microsoft Solutions Framework, http://msdn.microsoft.com/de-de/library/bb979125.aspx
10. Kittlaus, H.-B., Clough, P.N.: Software Product Management and Pricing: Key Success Factors for Software Organizations. Springer, Berlin (2009)
11. ISO/IEC 15288:2008. Systems and software engineering — System life cycle processes. International Standards Organization (2008)

# Applying ISO/IEC 12207:2008 with SCRUM and Agile Methods

Emanuel Irrazabal[1,2], Felipe Vásquez[2], Rafael Díaz[2], and Javier Garzás[1,2]

[1] KYBELE CONSULTING S.L.
{emanuel.irrazabal,javier.garzas}@kybeleconsulting.com
www.kybeleconsulting.com
[2] Kybele Research Group,
Universidad Rey Juan Carlos; Madrid, Spain
{emanuel.irrazabal,javier.garzas}@urjc.es,
{fa.vasquez,r.diazo}@alumnos.urjc.es

**Abstract.** Currently and in recent years several international initiatives specifically oriented to put together small and medium enterprises, processes and agile methods have been identified. Likewise, different studies have identified the mapping between agile methodologies and software development process models like CMMI-DEV and ISO/IEC 12207, but the studies related to ISO/IEC 12207 are based on the 1995 version. Therefore this work focuses on the relationship between agile practices, especially SCRUM, and a process subset from the 2008 version of the ISO/IEC 12207 standard. SCRUM is one of the most popular agile methods and is an incremental iterative process. These two characteristics mean dividing the project into phases or iterations and incremental delivery of the project. The relationships indicated in the work are obtained from the analysis of previous works and consulting experience at 25 enterprises that comply with the standard outcomes implementing agile methodologies. The main purpose of the study is to know the extent to which agile practices help in the implementation of practices indicated in this process model.

**Keywords:** ISO/IEC 12207:2008, SCRUM, agile.

## 1 Introduction

Increasingly, organizations agree the importance of control and improve software quality due to the impact it has on the final costs, as a distinctive from the competition and the image in customer-facing, especially taking into account the growth of software industry [1]. In this sense, the software process improvement is an activity that organizations want to implement in order to increase the quality and capability of their processes [2] and, consequently, the quality of their products and services.

However, numerous studies [3-5] show that application models like CMMI-DEV [6] are expensive in Small and medium enterprises (SME) and small development teams, their recommendations are complex to implement and the return of the investment occurs in the long term. Added to this are agile methodologies, a paradigm widely used by small development teams [7].

In this way, several international initiatives specifically oriented to put together SME, processes and agile methods have been identified [8-10]. Likewise, different studies [11-14] have identified the mapping between agile methodologies and software development process models such as CMMI-DEV or ISO/IEC 12207 [15]. In fact, mapping done between agile methods and ISO/IEC 12207 have been taken into account only the standard of year 1995. Therefore, this study shows the general way in which agile methods are related to SCRUM [16] by a subset of processes of ISO/IEC 12207:2008. For the choice of this subset of processes have been taken into account SCRUM characteristics and the related works that have served to make the subsequent analysis. Selected process areas were requirements management, project management, configuration management, measurement and life cycle management. The main goal has been to determine the extent to which agile practices help in best practices implementation proposed in the processes related to this key activities. To that end, comparisons have been made based on two sources: related works and authors' experience in consulting at 25 enterprises, mostly SME, that comply with the standard outcomes implementing agile methodologies

Section 2 shows other similar work, keeping in mind other process models. Section 3 justifies the subset of processes of ISO/IEC 12207:2008 chosen for this study. Also, section 4 details how the implementation of agile practices is aligned with best practices of the subset of chosen processes. Finally, section 5 proposes the final considerations and conclusions are made.

## 2 Related Work

Implementing ISO/IEC 12207:2008 has the drawback that is based on "what" but not "how" to guide processes; Agile methods can help to make the "how". Studies such as [11][13] deal with guidelines and mappings of Agile methods with respect to ISO/IEC 12207. while in [13] talks about that documentation required by the ISO/IEC 12207 is not in contradiction to agile philosophy but against over documentation that inhibit software development or developers work, Also mentions use of personal and tools to auto generate documentation from source code; while in [11] a mapping is made using agile methodologies practices like SCRUM, XP and Mobile-D method used in real projects; those mappings give mechanisms such as Sprint Planning, task status management and requirements definition activities for the implementation and testing activities of ISO/IEC 12207. However, both are based on the ISO ISO/IEC 12207:1995.

Recent studies, like [17] show the good match between Software Product Management (SPM) and agile methods like SCRUM. This is because their iterative nature and handling complex requirements are hard to manage in an agile environment. A case study was performed, showing the improvements of the Agile SPM based on SCRUM. However, there is evidence of only one specific case study performed on a specific company; also it needs to be proved against not complex requirements as well.

There are other studies on the mapping between the world of agile methods and processes such as CMMI or ISO 12207. Specifically [18] details the causes why there is a misconception about the union of both paradigms. The study details some of that causes which talk about aspects related to the different origins of CMMI, agile methods and terminology related causes, but it does not specify those causes about ISO 12207. The document affirms that paradigms can coexist because each of them supports different levels of a project. Therefore, as quoted above, CMMI is aimed to "what" and agile methods are aimed to "how".

Other studies such as [12] also investigate how difficult can be an implantation of an agile method in a changing environment due to many variables which make difficult to create appropriate measures to take into account the specifications of the environment. These investigations include study cases with implementation of agile methods in already established companies, such as [19]. This is a study that shows the implementation process of SCRUM in a company with a CMMI level 5 getting completely satisfactory results. These results were an increment of productivity as consequence of combining two paradigms. Again, this implantation takes into account CMMI and not ISO 12207.

## 3 ISO/IEC 12207 Processes Related with SCRUM

This mapping focuses on the ISO/IEC 12207:2008 standard, since this is currently widely used in industry and is the most relevant ISO standard in regard to software development. Previous works confirm this, although their have been focused on ISO/IEC 12207:1995. Therefore it was considered this work, taking into account the latest version of the standard.

ISO/IEC 12207:2008 defines a common framework for software life cycle processes, with well-defined terminology, that can be referenced by the software industry. The norm defines 43 processes that are to be applied during different activities in the software development process.

On the other hand, agile methods are a set of software development practices that respond to traditional methods, which emphasize the "engineering-based approach" [20]. In this regard, SCRUM is one of the most widely used methodologies. This methodology focuses on project management, with mechanisms for "empirical process control"; where feedback loops constitute the core element. Software is developed by in increments (sprints), starting with planning and ending with a review [16]. To continue the study has been made a selection of key areas that can meet SCRUM practices, in accordance with those used in previous comparisons with CMMI-DEV and ISO / IEC 12207:1995. The selected areas are: requirement management [20], project management [20] configuration management [13], measurement [12] and life cycle management [21]. Table 1 shows ISO/IEC 12207:2008 processes associated with selected practices.

**Table 1.** ISO/IEC 12207:2008 processes associated with SCRUM key areas

| SCRUM key areas | Process |
|---|---|
| Requirement management | Supply Process |
|  | Stakeholder Requirements Definition Process |
|  | System Requirements Analysis Process |
| Project management | Project Planning Process |
|  | Project Assessment and Control Process |
| Life Cycle management | Life Cycle Model Management Process |
| Configuration management | Configuration Management Process |
|  | Software Configuration Management Process |
| Measurement | Measurement Process |

## 4 Integrating ISO/IEC 12207 with SCRUM

After choosing the processes to be analyzed, it should review a core aspect to consider: difference between processes and the methodology that implements it. In Software Development, *what* and *how to do it*, have always been core aspects in process improvement; nevertheless, they have created confusion. Although several articles [18][22-24] are about the difficulty of linking norms like ISO/IEC 12207 with agile methods, the reality is that they are at different levels of abstraction [25]. Process models like ISO/IEC 12207, define best practices and *what* we expect to find in processes, but never shows *how to do it* [18][26-28], this is described in the methods. Therefore, the use of processes models and agile methods must not be considered a contradictory aspect but a complementary one.

Similar to as described [18] for CMMI, the main idea is that ISO/IC 12207:2008 is a process model, in other words, it not define specific processes, or procedures, or activities nor products that must be built to reach expected results (outcomes) for organization processes. They are good guidance practices that should be adapted to the company's needs (which could be taking place in different activities in projects life cycle), when implemented together, help to reach the outcomes of the process, for example, agile methods activities.

A problem that often happens when implementing agile methods is the lack of documentation. As stated in [29], "one of the principles in agile methods suggests that should not exist exhaustive documentation, which is correct, but problems appear when there is no documentation". The complete lack of any support documentation is not a good practice because it does not help to any directly involved staff in the development such as new developers or maintenance staff. In fact, agile processes need documentation too (for example, history of users, stacks of products or iteration, comments in code, reviews, progress charts, etc.). 12207:2008 doesn't require fully document in any case. The goal pursued is repeatability of activities, in other words, the knowledge is meant to be kept within the organization and only among their people.

As mentioned, ISO/IEC 12207:2008 does not put restrictions on the method to be used to implement the processes. Therefore, it is possible to implement these processes with agile methods as SCRUM. The study focuses on processes detailed in Table 1 and more deeply on project management area processes, since SCRUM mainly covers these areas [30]. Like says Dyba in [21], "The management of software

Project as long been a matter of interest. Agile methods have reinforced this interest, because many conventional ideas about management are challenged by such methods". Therefore *Project Planning Processes* and *Project Assessment and Control Process* of ISO/IEC 12207:2008 will be analyzed and the rest of selected processes will be studied in a more general way.

It have been defined a scale to carry out the study in a similar way to how other studies do it [31,32], this scale represents the degree of relationship between outcomes of ISO/IEC 12207:2008 and practices of SCRUM (Table 2). Each element of the scale has an associated percentage range.

**Table 2.** Coverage level Outcomes

| Outcome / Process coverage | Criterion | Associated percentage |
|---|---|---|
| Not Satisfied | The outcome is not implanted with SCRUM. | Between 0% and 49% |
| Partially Satisfied | The outcome is not fully implemented with SCRUM | Between 50% and 79% |
| Satisfied | The outcome is implemented fully with SCRUM. | Between 80% and 100% |

Once this first phase of the study is concluded, there will be calculated the degree of coverage of the *Project Planning, and Evaluation and Project Control*. This coverage degree is represented by a calculated percentage. This is calculated dividing the process outcomes: (1) satisfied, (2) partially satisfied, (3) not satisfied with SCRUM practices by the total number of outcomes in the corresponding process.

Finally, degree of implementation of nonspecific processes in management area of project is calculated.

## 4.1 Degree of Relationship with Project Planning Process

The purpose of Project Planning process is developing and communicating project plans, in an effective and viable way [15]. This process consists of 6 outcomes.

| Outcome 1. The scope of the work for the project is defined | | Satisfied |
|---|---|---|
| Goal | Establish a high-level structure which identifies the project deliverables. This structure is called "Work Breakdown Structure" (WBS). | |
| SCRUM Practices | The project scope initial definition is done during the initial phase when interested parties (stakeholders) participate in the creation of the product stack (Product Backlog). In this case, EDT consists of the product stack and all iterations set (Sprints) predefined. Estimates of detailed tasks are made at the start of each iteration (Iteration Planning Meeting). | |

| Outcome 2. The feasibility of achieving the goals of the project with available resources and constraints are evaluated | | Satisfied |
|---|---|---|
| Goal | Align project objectives with available resources and estimated, taking into account existing constraints. | |
| SCRUM Practices | The viability analysis is performed during the iteration planning meeting and during each iteration follow-up meeting. At each planning meeting, the team, the Product Owner and SCRUM Master define features that can be developed in the iteration, and at each follow-up meeting the team indicates if it has been found or it is planned to find any impediment or restriction. | |

| Outcome 3. The tasks and resources necessary to complete the work are sized and estimated. | | Satisfied |
|---|---|---|
| Goal | Estimate the project effort and cost based on a defined method. | |
| SCRUM Practices | The estimate is made on two levels: product stack (Product Backlog) and stack iteration (Sprint Backlog). Estimates of the product stack are very approximate, are realized only when it needs to be done during the iteration. To conduct the meetings in which the team estimates the effort and duration of tasks, using methods such as the "Poker Estimate" [33] or variants such as the "Fibonacci Sequence". The estimated cost of the project is not explicitly treated. | |

| Outcome 4. Interfaces between elements in the project, and with other project and organizational units, are identified. | | Not Satisfied |
|---|---|---|
| Goal | Identify human and material resources that are shared on the project with other projects and / or with the same organization. | |
| SCRUM Practices | SCRUM methodology does not refer explicitly to the identification of possible relationships between projects, other projects and organizational units (eg: load resources, organizational plans, etc.). | |

| Outcome 5. Plans for the execution of the project are developed. | | Partially Satisfied |
|---|---|---|
| Goal | Make a planning project that contains descriptions of activities and associated tasks, where is specified the calendar, tasks, estimates, resources, risks and budget. | |
| SCRUM Practices | The minimum plan to start a SCRUM project is to define the vision and scope of the project and product stack [34]. The scope describes why the project is being conducted and what the final state to be achieved. The product stack defines the functional and nonfunctional requirements (prioritized and estimated) that the system must meet to achieve the defined scope. Also defines iterations that compose the project.<br>In the planning meeting of the iteration, the iteration stack is made, which describes the functionality that will develop in the iteration, assigning each task to a person, and indicating the estimate associated with each person.<br>To show in a simple way the product development general plan, and expected developments, the "Burn-up chart" is developed, which presents the planned product releases, functionality of each one, estimated speed, likely dates for each version, expected error in the estimates, and real progress [33].<br>The risks are discussed in the follow-up meetings, where among others, the team indicates whether found or are expected to find with any impediment to the development of tasks.<br>The planning of the project cost is not treated in an explicit way. | |

| Outcome 6. Plans for the execution of the project are activated | | Satisfied |
|---|---|---|
| Goal | Communicate and obtain authorization to carry out the project. | |
| SCRUM Practices | The launch of each new increment of functionality is made in iteration's planning meeting. It is a meeting conducted by the Scrum Manager, which should assist the product owner and the entire team and other stakeholders, where features to be developed in that iteration are presented and planned. | |

### 4.2 Degree of Relationship with Project Assessment and Control Process

The purpose of the Project Assessment and Control Process is to determine the status of the project and ensure that is done in accordance with the plans and schedule, planned budgets and meeting the technical objectives [15]. This process consists of 4 outcomes.

| Outcome 1. Progress of the project is monitored and reported | | Satisfied |
|---|---|---|
| Goal | Track the actual project progress compared to planned. | |
| SCRUM Practices | Project control is performed during the daily follow-up meetings and review of the iteration.<br>In daily tracking meetings each team member comments on the tasks they are working, if it has founded or will expect to meet with an impediment, and updates the stack of completed tasks. At these meetings progress chart of the iteration is used (Sprint Burn-down) which shows daily team's speed and progress of its activities related to iteration. The review meeting of the iteration is performed at the end of the iteration in which the team presents the increment built in the iteration and provides information on the progress of the product. At these meetings progress charts are used (Product Burn-down), which shows the speed with which the team implements the elements of the product's stack (Product Backlog). This helps to track the implementation of the functions and provides some visibility into their progress and need for re planning. | |

| Outcome 2. Interfaces between elements in the project, and with other project and organizational units, are monitored | | Not Satisfied |
|---|---|---|
| Goal | Control human and material resources that are shared on the project with other projects and / or with the same organization. | |
| SCRUM Practices | As with the outcome 2 of the process "Project Planning" SCRUM methodology does not refer explicitly to a control implementation of possible relationships between the projects, other projects and organizational units (eg: Loads of resources, organizational plans, etc.). | |

| Outcome 3. Actions to correct deviations from the plan and to prevent recurrence of problems identified in the project are taken when project targets are not achieved | | Satisfied |
|---|---|---|
| Goal | Collect and analyze deviations that occur in the implementation of the project regarding the planning and determine corrective actions needed to resolve them. | |
| SCRUM Practices | Burn-down charts provide support for planning corrective actions, providing timely information on the degree of deviation between the actual and planned situation. In addition, at daily tracking meetings the team reports all impediments, to be recorded on a whiteboard or a list of impediments. The SCRUM Master is responsible to solve impediments, as soon as possible, taking corrective action. | |

| Outcome 4. Project objectives are achieved and recorded | | Satisfied |
|---|---|---|
| Goal | Collect and analyze deviations that occur in the implementation of the project regarding the planning and determine corrective actions to resolve them. | |
| SCRUM Practices | The information of the work involved in the project is kept updated in both the product stack and the iteration stack. Every day, each team member updated the product stack, time remaining to the tasks in which is working, until they are completed and time remaining to 0 pending.<br>After each iteration, the product stack is updated with features developed and carried out a review meeting of the iteration, where the team presents to the product owner and other stakeholders the increase built in the iteration given the initial target, features list that are included and those that have been developed. | |

In summary, Figure 2 shows coverage degree of project management processes with SCRUM practices.

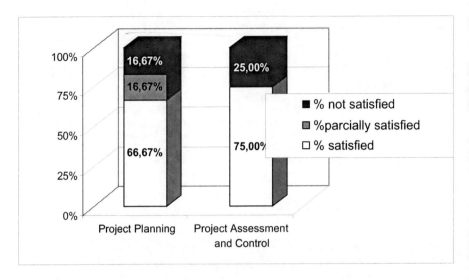

**Fig. 1.** Coverage degree of project management processes using SCRUM

## 4.3 Degree of Relationship with the Other Selected Processes

In this section are presented the remaining processes of the model ISO / IEC 12207:2008 selected in Table 1 that are not directly related to the principles of project management described SCRUM but for which there is some degree of coverage.

| Supply process [P1]<br>Stakeholder Requirements Definition Process [P2]<br>System Requirements Analysis Process [P3] || **Partially Satisfied** |
|---|---|---|
| Goal | [P1]: Give the customer a product that meets the requirements.<br>[P2]: Define the needed requirements for the system to provide needed services in a defined environment.<br>[P3]: Transforming the requirements of "stakeholders" in a set of system technical requirements that will guide the design of the system. ||
| SCRUM Practices | The relationship between customers, development team, development tasks and software product is made from the beginning and throughout the project. Requirements are expressed by customers in the product stack, which is a list of functional and nonfunctional requirements prioritized according to customer needs.<br>In each iteration, customer requirements that have to be developed will be refined to get the system requirements. This will ensure traceability between customer needs and system requirements.<br>In the iteration reviews it is showed the progress of the project and validated customer deliveries. All involved in the project and especially the customer must provide feedback on the iteration outcome, which is always a new completed product functionality. ||

| Configuration Management Process<br>Software Configuration Management Process || **Partially Satisfied** |
|---|---|---|
| Goal | Establish and maintain the integrity of all work products and items that compose the software product in a project or process and make it available to interested parties. ||
| SCRUM Practices | On Teams where the coding is done concurrently, it is necessary to carry out activities that enable effective collaborative work. Although not mentioned directly, SCRUM practices often require a strategy for software configuration management to work ||

properly and to maintain good practice such as is shown in ISO/IEC 12207:2008 model. Among SCRUM practices is found the practice related to the functionality delivery at the end of each iteration. A precondition for this is to get that the main development does not contain components not fully developed, and ensure integration and consistency. Maintain continuous integration, define and use a branch policy and continuous evaluations are common practices in an agile environment.

| Life cycle model management Process | Partially Satisfied |
|---|---|
| Goal | Define, maintain and ensure the availability of policies, processes and life cycle models, to be used by the organization. |
| SCRUM Practices | SCRUM is based on continuous adaptation to the circumstances of the project's evolution. Uses iterative and incremental life cycle model [36]. To ensure the proper functioning of the methodology in the organization, it is established the SCRUM Master role. However, it is not explaining how to define, manage, and improve their methodology throughout the project. Proper documentation that describes the steps that the organization follows when using SCRUM would be one of the key aspects to comply with this process. |

| Measurement process | Partially Satisfied |
|---|---|
| Goal | Collect, analyze and report about data relating to developed products and implemented processes within the organizational unit, to support effective processes management and demonstrate objectively the quality of products. |
| SCRUM Practices | Information needs in agile methodologies focus on visibility of project status. In [34] mention the typical needs of information from the standpoint of all involved in Scrum: the team, SCRUM Master, customers and the management team of the organization, which reviews all possible measures to meet their needs: number of impediments by daily meeting, team members satisfaction, code size, number of errors during the iteration review meeting, tasks completed during the review, etc.
SCRUM does not define mechanisms to carry out the measurement tasks, however, information needs can be met without additional activities [34], recording the information normally obtained in the meetings and revisions.
An example can be found in [35]. This paper describes the experience and the approach used in an agile IT company to design and initiate a measurement program taking into account the specifities of their agile environment, principles and values. |

## 5 Final Considerations and Conclusions

This article describes the coverage degree among a subset of ISO/IEC 12207:2008 process, agile methodologies and SCRUM. We have focused on the subset of processes in accordance with comparisons related to other process models or older ISO/IEC 12207 versions. Comparisons have been made based on two sources: related works and authors' experience in consulting at enterprises, mostly SME, that comply with the standard outcomes implementing agile methodologies. In brief, from a sample of 25 organizations 56% used SCRUM to address the area of project management (see Figure 3).

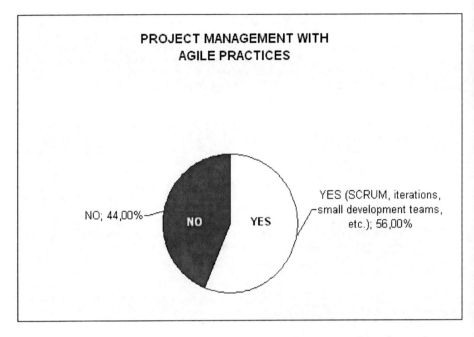

**Fig. 2.** Percentage of companies that meet the project management with agile practices

This result is supported from the analysis in Section 4. See Figure 1 it is concluded that the implementation of SCRUM would be reached practically 83% of the Project Planning process and 75% of the Project Assessment and Control Process. Similarly, other 7 processes can be partially implemented with agile practices, according with Section 4.3. It should be recalled that the use of a methodology does not ensures the direct fulfillment of standards. Yet it has been demonstrated that agile practices can help with compliance, as any best practice. For future work is planned to conduct a more detailed study of the rest of the processes identified in Table 1. And we will also contrasting analysis results with expertise in enterprises.

**Acknowledgments.** This work has been partially funded by MODEL-CAOS project (TIN2008-03582/TIN) financed by the Spanish Ministry of Education and Science (TIN2005-00010/).

# References

1. Piattini, M., Garzás, J.: Fábricas de software: Experiencias, tecnologías y organización. Ra-ma, Paracuellos del Jarama (2010)
2. Hurtado, J., Pino, F., Vidal, y.J.: Software Process Improvement Integral Model: Agile SPI. In: T.R.S.-S.-O.A.-R.-.-V., Universidad del Cauca - Colciencias, Popayn (2006)
3. Hareton, L., Terence, Y.: A Process Framework for Small Projects. Software Process Improvement and Practice 6, 67–83 (2001)

4. Saiedian, H., Carr, N.: Characterizing a software process maturity model for small organizations. ACM SIGICE Bulletin 23(1), 2–11 (1997)
5. Staples, M., et al.: An exploratory study of why organizations do not adopt CMMI. Journal of Systems and Software 80(6), 883–895 (2007)
6. CMMI Product Team, CMMI for Development, Version 1.3, Technical Report CMU/SEI-2010-TR-033, Carnegie Mellon University/Software Engineering Institute (2010)
7. Coleman, G., O'Connor, R.: Investigating software process in practice: A grounded theory perspective. J. Syst. Softw 81(5), 772–784 (2008)
8. Ramachandran, M.: A Process Improvement Framework for XP Based SMEs. In: Baumeister, H., Marchesi, M., Holcombe, M. (eds.) XP 2005. LNCS, vol. 3556, pp. 202–205. Springer, Heidelberg (2005)
9. Nawrocki, J., et al.: Toward maturity model for extreme programming. In: 27th Euromicro Conference, pp. 233–239 (2001)
10. McCafferry, F., et al.: Ahaa - Agile, Hybrid Assessment Method for Automotive, Safety Critical SMEs. In: 30th Int. Conference on Software Engineering, pp. 551–560 (2008)
11. Pikkarainen, M.: Mapping agile software development onto ISO 12207. Technical Report (2006),
    http://www.agile-itea.org/public/deliverables/ITEA-AGILE-D2.9_v1.0.pdf
12. Ktata, O., Ghislain, L.: Designing and implementing a measurement program for Scrum teams: what do agile developers really need and want? In: Proceedings of the Third C* Conference on Computer Science and Software Engineering (C3S2E 2010), pp. 101–107. ACM Press, New York (2010)
13. Theunissen, W.H.M., Kourie, D.G., Watson, B.W.: Standards and agile software development. In: SAICSIT 2003: Proceedings of the, annual research conference of the South African institute of computer scientists and information technologists on Enablement through technology (Republic of South Africa), pp. 178–188 (2003)
14. Marcal, A., Furtado Soares, F., Belchior, A.: Mapping CMMI Project Management Process Areas to SCRUM Practices. In: Proceedings of the 31st IEEE Software Engineering Workshop (SEW 2007), pp. 13–22. IEEE Computer Society, Washington (2007)
15. ISO, ISO/IEC 12207:2008. Systems and software engineering - Software life cycle processes. International Organization for Standarization (2008)
16. Schwaber, K., Beedle, M.: Agile software development with Scrum. PrenticeHall PTR, Upper Saddle River (2001)
17. Vlaanderen, K., Jansen, S., Brinkkemper, S., Jaspers, E.: The agile requirements refinery: Applying SCRUM principles to software product management. Inf. Softw. Technol. 53(1), 58–70 (2011)
18. Glazer, H., et al.: CMMI or Agile: Why not Embrace Both! Technical Note CMU/SEI-2008-TN-003. Software Engineering Institute (2008),
    http://www.sei.cmu.edu/reports/08tn003.pdf
19. Jakobsen, C.R., Sutherland, J.: Scrum and CMMI Going from Good to Great. IEEE, Washington (2009)
20. Dyba, T.: Improvisation in Small Software Organizations. IEEE Softw. 17(5), 82–87 (2000)
21. Tore Dyba, T., Dingsayr: Empirical studies of agile software development: A systematic review. Inf. Softw. Technol. 50(9–10), 833–859 (2008)
22. James, B.: The Hard Road From Methods to Practice. Computer 30(2), 129–130 (1997)
23. James, B.: Reconcilable differences. Softw. Dev. 5(3), 34–40 (1997)

24. James, B.: Enough About Process: What We Need are Heroes. IEEE Softw. 12(2), 96–98 (1995)
25. Navarro, J.M.: Experiencia en la implantación de CMMI-DEV v1.2 en una micropyme con metodologías ágiles y software libre. Revistas española de innovación, calidad e ingeniería del software 6(1), 6–15 (2010)
26. Wang, Y., King, G.: Software Engineering Processes: Principles and Applications. CRC Press, Boca Raton (2000)
27. Osterweil, L.: Software Processes Are Software Too. In: Proceedings of the 9th International Conference on Software Engineering (1987)
28. Yingxu, W., Antony, B.: Process-Based Software Engineering: Building the Infrastructures. Ann. Softw. Eng. 14(1-4), 9–37 (2002)
29. Selic, B.: Agile Documentation, Anyone? IEEE Softw., vol. 26(6), pp. 11–12 (2009)
30. Schwaber, K., Sutherland, J.: The Scrum Guide. In: Scrum Alliance (2010), http://www.scrum.org/storage/scrumguides/Scrum%20Guide.pdf#view=fit
31. Marcal, A.S.C., Soares, F.S., Belchior, A.D.: Mapping CMMI Project Management Process Areas to SCRUM Practices. In: SEW 2007: Proceedings of the 31st IEEE Software Engineering Workshop. IEEE Computer Society Press, Los Alamitos (2007)
32. Turner, R., Jain, A.: Agile Meets CMMI: Culture Clash or Common Cause? In: Proceedings of the Second XP Universe and First Agile Universe Conference on Extreme Programming and Agile Methods - XP/Agile Universe. Springer, Heidelberg (2002)
33. Cohn, M.: Agile Estimating and Planning, pp. 49-60. Prentice Hall, Massachusetts (2005)
34. Viljan, M., Natasa, Z.: Measurement repository for Scrum-based software development process. In: Proceedings of the 2nd WSEAS International Conference on Computer Engineering and Applications. World Scientific and Engineering Academy and Society (WSEAS), Acapulco (2008)
35. Ktata, O., Lévesque, G.: Designing and implementing a measurement program for Scrum teams: what do agile developers really need and want? In: Proceedings of the Third C* Conference on Computer Science and Software Engineering (C3S2E 2010), pp. 101–107. ACM Press, New York (2010)
36. Sutherland, J.: Agile Development: Lessons Learned From The First Scrum Internet (October 2004), http://www.scrumalliance.org/resource_download/35, (February 20, 2011)

# Agile and SPICE Capability Levels

Celestina Bianco

Systelab Technologies
`Celestina.bianco@systelabsw.com`

**Abstract.** The paper presents the results of an exercise that considers the principles of the Agile Manifesto and of the Declaration of Interdependence, and evaluate how much their implementation is compatible with a high target profile in a Spice Assessment.

**Keywords:** Agile, Capability, Assessment.

## 1 Introduction

The Agile Manifesto was written as a conclusion of research into better ways of developing software and helping others do it. The Manifesto reads as:

    Individuals and interactions  *over*  processes and tools
    Working software  *over*  comprehensive documentation
    Customer collaboration  *over*  contract negotiation

There is a "lazy" interpretation of the manifesto, that replace the word "over" with "instead", and Agile can be presented as document-less, test-less, rule-less, etc; that down-sizing of the approach is not compatible with SPICE nor with any other maturity model. Other interpretations brought to develop methodologies that are widely used as SCRUM, XP, Lean Driven, and others. All of them have their "decalogue", all have risk management, test and customer validation and acceptance as a major value.

## 2 Agile and SPICE

Let´s consider the expected result if I perform an assessment on a company that is "Agile":

- They have artifacts – customer requirements, test, **working software**, ..
- They have engineering processes that are repeatable and similar company-wide – the rules of the **methodology** and the customization done in the Company

The assessment will confirm that the Company has the capability to satisfy the customer needs, in fact improving customer satisfaction was the reason to adopt Agile. On the contrary, the assessment will probably be useful in detecting one of the many mis-use or mis-labeling of Agile, with short cuts that forget the basic concepts and goals of the methodology.

From the quick analysis above, it appears that an Agile company will easily have target profile 2 for the engineering and related processes, whereas a higher level is "incompatible" with the Agile priorities (intrinsic in the *"over"* statement) because of the formalization of the process and metrics of SPICE Level 3 and higher.

- Can it imply that a Company at level 3 or higher will have resistance in adopting Agile? Or
- Will Agile be in conflict with requirements in critical fields where level 3 is expected?

## 3 SCRUM: An Agile Methodology

If we take as example one of the more popular Agile Methodologies, SCRUM, its basic concepts can be summarized in the following table:

**Table 1.** SCRUM

| | |
|---|---|
| Scrum | An iterative, incremental process for developing software in dynamic environments. Scrum consists of a series of 10 to 30 day sprints, each sprint producing a releasable product |
| Backlog | A prioritized list of all work to be completed prior to releasing a product; only one person maintains and prioritizes the backlog list. Each element is a "user story". An initial general Product Backlog is refined in Sprint backlogs. A backlog includes both functional and no functional requirements for the product. The description of the functionality is the minimum needed to enable development and testing; additional information can be added during the development.<br>A backlog is mandatory, other documents maybe issued as a result of the implementation of a user story. |
| Product Owner | The person who listen to and interacts with all stakeholders –including customer, financial, layer, teams- , defines priorities and characteristics of the product described in the backlog. Provides details of requirements for each Sprint |
| Scrum Master | Handles the process inside the project, responsible to coordinate the team and above all to remove impediments to comply with goals and time frame |
| Team | Group implementing the product, includes architects, analysts, developers, testers. 5 to 9 people that work at a stable peace, fully responsible for the part developed and committed to the project |
| Sprint | A short burst of work lasting no more than 30 days during which a releasable product and other deliverables are built, as indicated by the sprint backlog.<br>• Every sprint has a specific goal.<br>• A releasable product demonstrating the goal will be completed by the team during the sprint.<br>• Once the sprint is underway, new backlog cannot be added to the sprint<br>• A kick-off meeting is held to explain the goal, and detail user histories, followed by a planning meeting to estimate the efforts<br>• Frequent check-points are established<br>• The Sprint if finalized with a meeting of presentation to the product owner and stakeholders by members of the team and with a retrospective evaluation of general performance and process issues |
| Scrum daily meeting | • Meeting conducted daily by the team<br>• The meetings don't last for more than 30 minutes.<br>• Conversation is restricted to the team members answering 3 questions:<br>  1. What have you done since the last scrum meeting?<br>  2. What has impeded your work?<br>  3. What do you plan on doing between now and the next scrum meeting?<br>• To resolve impediments, the scrum master can make decisions immediately, or define an action plan |

The methodology is able to cover most of the Engineering Process Group, and part of others. The set of processes that directly or indirectly are covered or needed for a proper roll-out and execution of Agile are in the following table:

**Table 2.** Processes covered by a proper roll-out and execution of Agile

| Group | Processes | Notes |
|---|---|---|
| Supply (SPL) | *Product release*<br>*Product acceptance support* | By the product owner |
| Engineering (ENG) | Requirements<br>*Software requirements analysis*<br>*Software design*<br>*Software construction*<br>*Software integration*<br>*Software testing*<br>*System integration*<br>*System testing* | Core of the methodology, system level managed by the product owner |
| Resource and Infrastructure (RIN) | *Human resource management*<br>*Training*<br>*Infrastructure* | Intrinsic request of the method |
| Management (MAN) | Organizational alignment<br>Organizational management<br>*Project management* | Agile is designed to accomplish with business goal |
| Support (SUP) | Verification<br>*Validation*<br>*Joint review*<br>Product evaluation<br>Documentation<br>*Configuration management*<br>~~Problem resolution management~~<br>*Change request management* | The scope of documentation is not as expected by other methodologies<br><br>Sup-9, Problem Resolution Management, can be seen as part of the change management, defects are items of the next sprint backlog |
| Process Improvement | PIM.1 Process establishment<br>PIM.2 Process assessment<br>PIM.3 Process improvement | Agile is rolled-out with the purpose of Process Improvement and the process established need learning curve and the efficacy is continually monitored |

# 4 Capability Assessment

The processes that are surely part of the profile of any company that properly establish an Agile development process, and properly follow it are marked in blue. If we focus on those processes, which rating of the Capability Attributes can be reached if properly defined and implemented?

If correctly implemented SCRUM will Fully fulfill Level 2 capability attributes.

| Process Performance | Fully (> 85) |
|---|---|
| Performance Management | Fully |
| Work Product Management | Fully |

Level 3 requires Capability Attributes to be fulfilled completely for Level 1 and 2, largely at Level 3.

| Process Definition | Largely or fully > 50 |
|---|---|
| Process Deployment | Largely or fully > 50 |

We will analyze in detail fulfillment at level 3:

**Table 3.** Mapping Level 3 on SCRUM

| Attribute | Description | Scrum rate |
|---|---|---|
| PA 3.1 | a) a standard process, including appropriate tailoring guidelines, is defined that describes the fundamental elements that must be incorporated into a defined process; *(documentation Control is to be fully established to reach Fully)* → | Largely |
|  | b) the sequence and interaction of the standard process with other processes is determined; | Largely |
|  | c) required competencies and roles for performing a process are identified as part of the standard process; | Fully |
|  | d) required infrastructure and work environment for performing a process are identified as part of the standard process; | Largely |
|  | e) Suitable methods for monitoring the effectiveness and suitability of the process are determined. | Fully |
| PA 3.2 | a) a defined process is deployed based upon an appropriately selected and/or tailored standard process; | Largely |
|  | b) required roles, responsibilities and authorities for performing the defined process are assigned and communicated; | Fully |
|  | c) personnel performing the defined process are competent on the basis of appropriate education, training, and experience; | Fully |
|  | d) required resources and information necessary for performing the defined process are made available, allocated and used; | Fully |
|  | e) required infrastructure and work environment for performing the defined process are made available, managed and maintained; | Fully |
|  | f) appropriate data are, collected and analysed as a basis for understanding the behavior of, and to demonstrate the suitability and effectiveness of the process, and to evaluate where continuous improvement of the process can be made | Largely |

**Table 4.** Declaration of Interdependence

| Declaration of Interdependence | Comments |
|---|---|
| We **increase return on investment** by making continuous flow of value our focus. | ROI Needed more than a contract to provide and guarantee continuous service to the customer; the declaration claims that ROI is re-invested in value for customers |
| We **deliver reliable results** by engaging customers in frequent interactions and shared ownership. | The goal of Engineering process in critical systems is to achieve and guarantee reliability |
| We **expect uncertainty** and manage for it through iterations, anticipation, and adaptation. | Even if uncertainty is not the most relevant characteristic of critical systems that require a Level 3, adaptation is mandatory to achieve business objectives |
| We **unleash creativity and innovation** by recognizing that individuals are the ultimate source of value, and creating an environment where they can make a difference. | Complex systems generally deal with innovation and multidiscipline. The agile management of interaction of experts with engineers avoids risks deriving from misunderstanding of real specifications |
| We **boost performance** through group accountability for results and shared responsibility for team effectiveness. | Commitment of each participant in a project makes systems more robust and is a warranty for the capability of a company |
| We **improve effectiveness and reliability** through situationally specific strategies, processes and practices. | In an Agile context strategies to ensure effectiveness and reliability are not static, but not uncontrolled |

Maturity Capability at Level 3 is often asked of companies that develop complex and critical products. The Declaration of Interdependence, published in 2005, that inspired the Agile Project Management, augment capability, even if a direct mapping of the statement of the declaration to the Capability Attribute is difficult.

## 5 Conclusions

We have seen that – in spite of a huge difference that might be expected- goals of Agile and SPICE are coincident in some degree, as both target process improvement.

Agile development and auxiliary processes needed to support it, if correctly implemented, can allow a company to achieve capability level 3. In addition, in some companies either for law, business or regulatory requirements, the pure form is extended with additional processes, tasks and outcomes, to comply with standards and generally the Software design methodology adopted is used together with a specific and formalized Quality System. These additional tasks are those that would increase a full capability at Level 3, in its strict conventional way.

*We leave a question open: is the concept and goal of achieving Capability Level 4 compatible with the goals of a company that has chosen to establish an Agile process?*

Contrary to what might be initially be considered, an Agile company can get renewed motivation from a SPICE assessment. On the other hand: in a Capability evaluation of companies that adopt Agile, can an assessor adopt an "agile" approach as well, and be compliant with ISO 15504? For example: a company that develops Software with an Agile approach, will probably have the rest of the activities organized on the same principles, they will be rolled out and have people trained. In that context, can the whole set of training material be considered a formalization of the process? Can retrospective meeting analysis be considered as metrics for improvement?

## References

1. Agile Manifesto, http://agilemanifesto.org/
2. Declaration of Interdependence, http://pmdoi.org/
3. ISO/IEC FDIS 15504-2. Software engineering — Process assessment — Part 2: Performing an Assessment. Part 5: An exemplar Process Assessment Model
4. Kniberg, H.: Scrum and XP from the Trenches. InfoQ Enterprise Software Development Series

# Application of Lean Principles in Automotive Software Projects

Smitha Bhandary and Shah Quadri

Wipro Technologies, Bangalore, India
{smitha.saikumar,shah.quadr}@wipro.com
http://www.wipro.com

**Abstract.** Automotive software industry today is more globalized due to availability of low cost domain specific service providers, and this is leading to growth of supply chain from Tier-1 to Tier-2 and so on. Accordingly there are numerous challenges in coordinating this supply chain. Any obstacles to the flow of information will result in delays thereby having an impact on time to market and cost

In this paper we present the actual case study of application of lean principles in an automotive software maintenance project and how it has helped in achieving an efficient system by improving customer value and reducing waste.

**Keywords:** Lean, Automotive Software, Maintenance project, Case Study.

## 1 Introduction

Automobiles today are different in that they present innovations in communication, mobility, safety and entertainment. Today Automotive Industry is seeing significant demand for innovative and efficient vehicles. In Automobiles it is estimated that 90% of innovation by 2012 will be electronics related and 80% of that in the areas of software. With today's globalized software development; adopting Lean principles reduces delays, minimizes waste, maximizes value and reduces cost. Already Lean principles in manufacturing industry have done wonders in achieving the following:

- Reduction in cost
- Reduced cycle time to market
- Improvement in quality

This paper explains how Lean techniques have benefitted an Automotive software maintenance project[3].

## 2 Background

Lean thinking is derived from the lean production system theory. In 1996, Professor James P. Womack from Massachusetts Institute of Technology published the "Lean thinking" which summed up the lean production principles and further elaborated the idea of Lean production [1]. Our case study is based on the below three core Lean principles:

1. Understanding Customer Value- Only what customers perceive as value is important.
2. Value Stream Analysis –. Operational efficiency opportunities are unearthed by Value Stream Mapping (VSM) in any business processes.
3. Perfection – This process of continuous improvement opens up methods to reduce cycle time, improve quality and increase productivity [1] as mentioned by standards such as Automotive SPICE [2].

During the project executionfollowing were the major challenges:

1. Defect fix productivity was below customer expectations
2. Customer wanted hands-free execution at offshore by reducing dependency on their team
3. Process was found to be too sequential with lots of redundancy.

# 3 Case Study

In order to manage the challenges Lean principles were applied on the project. Though Lean provides plethora of tenets, three tenets namely Value Stream Mapping, Workload leveling and Visual Controls were selected and applied in this project for duration of 5 months.

| Lean Tenet | Reason for Selection |
|---|---|
| Value Stream Mapping | To understand defect fixing process and identify areas for improvements |
| Workload Levelling | To optimize the balancing of work. |
| Visual Controls | To improve planning and monitoring through visual aids |

## 3.1 Value Stream Mapping

This technique was used to identify steps which had no value and could be immediately avoidable. The project applied value stream mapping on the defect fixing life cycle the entire set of activities of the defect life cycle was examined by identifying Actual Time (AT) and Value Added Time (VAT) for each activity the Overall Process Efficiency Factor (PEF) was calculated as shown below:

|  | Team-1 Analysis | | Team-2 Analysis | |
| :---: | :---: | :---: | :---: | :---: |
| Activity | AT (Hr) | VAT (Hr) | AT (Hr) | VAT (Hr) |
| Activity-1 | 0 | 0 | 0 | 0 |
| Activity-2 | 6 | 2 | 6 | 2 |
| Activity-N | 0.25 | 0.25 | 0.25 | 0.25 |
| Number of Engineers | 4 | | 16 | |
| PEF | 33% | | 34% | |
| Weighted PEF | 33.50% | | | |

Process Efficiency Factor (PEF) = $\Sigma$ (VAT)/$\Sigma$ (AT) represented as percentage (%).

Through the analysis it was found that the current PEF was at 33.5% and required immediate focus. The following plan was followed which helped the team to reduce the Non-value add activities drastically

1. Reducing the dependency on customer by strengthening technical competency of local team at Wipro and this was measured using dependency factor
2. Avoiding sequential process and working in parallel on more than one defect(s).

### 3.2 Workload Leveling

Demand variations cause workload fluctuations necessitating an effective workload balance (Heijunka) strategy. It was found that 30% of Wipro team was working on 10% of defects due to low inflow while 25% team was working over-time on 60% of the defects due to high inflow of work. Each module was analyzed for amount of inflow, resource loading and productivity. The productivity factor was considered to fetch faster results. Based on the above analysis, 3 engineers were shuffled between the modules which helped to improve productivity by almost 15%.

| Module | In-Flow of work | # resources assigned | Productivity | Leveling Plan |
| :---: | :---: | :---: | :---: | :---: |
| Module-1 | High | 1 | Satisfactory | Increase Team Size |
| Module-2 | High | 1.5 | Improvement req. | No Change |
| Module-n | Low | 2 | Low | Reduce Team Size |

### 3.3 Visual Control

Visual controls were deployed to showcase the plan for current week and also project the performance of each team member during the past week. The advantages of using the Visual controls for tracking were:

1. Team got view of assignments to all.
2. This resulted in immediate actions on issues, assignments, quick feedback to improve continuously thereby delivering value to customer.

### 3.4 Benefits

1. At the end of fourth months Productivity and Process Efficiency Factors were compared. With the application of Lean tenets, clubbed with improvement in competency, an overall improvement of 95% in productivity and 48.80% improvement in PEF. Below are the results:

| Month | Defect Productivity | PEF | Waiting Time Saved |
|---|---|---|---|
| Month-1 | 1.43 | 33.50% | - |
| Month-2 | 2.27 | 40.40% | 160 hours |
| Month-3 | 2.34 | 43.60% | 270 hours |
| Month-4 | 2.79 | 48.80% | 171 Hours |

## 4 Conclusions

In this paper we have presented an ongoing experience of application of Lean principles and few Lean tenets on automotive software maintenance projects. By adopting Lean principles, we noticed overall it enhances the efficiency of the system, reduces cost and improves cycle time and quality and delivers value to the customer.

## References

1. Womack, J.P., Jones, D.T.: Lean Thinking. Banish Waste and Create Wealth in Your Corporation
2. Automotive SIG. Automotive SPICE© Process Assessment Model (v2.5) and Process Reference Model (V4.5) (2010)
3. Lientz, B.P., Swanson, E.B.: Software Maintenance Management: A Study of the Maintenance of Computer Application Software in 487 Data Processing Organizations. Addison-Wesley, Reading (1980)

# Experiences Developing TMMi® as a Public Model

Matthias Rasking

Accenture GmbH, Kronberg, Germany and TMMi Foundation, Dublin, Ireland
matthias.rasking@accenture.com

**Abstract.** Over the years, there have been many models developed to assess Test and Quality related processes, which have not gained any universal acceptance by the industry at large. Even today, there is no fully accepted, independent standard model available. This paper describes the experiences of the TMMi Foundation with developing an open, independent and public standard model for Test and Quality related processes called the Test Maturity Model Integration. Instead of relying on a proprietary method and so-called "best practices" from individuals, the TMMi Foundation set out to develop a model for testing and quality-related processes that captures the know-how and experiences of test practitioners, related process models. TMMi is emerging as the industry standard developed by practitioners for practitioners.

**Keywords:** Model Development, CMMI, TMMi, SPICE.

## 1 Introduction

Over the years, there have been many models developed to assess the Test and Quality related processes, which have not gained any universal acceptance by the industry at large[1]. While many models exist even today, there is no accepted, standard model available[2]. Therefore in 2005 a group of leading test and quality practitioners from across the world decided to define a model and launched the TMMi Foundation. This non-profit organization has the objective of promoting and providing the infrastructure to manage a standard for a testing-specific maturity model available to all – the Test Maturity Model Integration (TMMi).

Working across industry and academic boundaries to develop a standard Model that can be used in isolation or in support of other process improvement models such as CMMI® is never easy considering the sometimes conflicting objectives of each group. Moreover, with Testing Services enjoying an unprecedented growth story in the Software Development and Computer Services industry[3], a lot of pressure exists to get the model right and universally accepted by the industry, covering all sectors, applications and delivery/service models.

Previous models mostly failed to get the required universal support: usually because they were not being based on the needs of a broad enough stakeholder group and not consistently maintained and improved to keep pace with emerging trends and needs. Moreover, alignment with existing industry models was sometimes not at the level necessary to show the success the combination of different models can have.

The TMMi Foundation aims to solve these issues with model development for testing and quality-related processes. This paper focuses on the process of model development by sharing lessons learned when involving multiple industry groups and reviewers as well as the challenge of marketing a new, public model in a mature market.

## 2 Background

TMMi as a Process Reference Model (PRM) was based on the TMM framework[4] as developed by the Illinois Institute of Technology as one of its major sources. In addition to the TMM, it was influenced by the work done on the Capability Maturity Model Integration (CMMI), a process improvement model that has widespread support in the IT industry as well as quality models from Gelperin and Hetzel[5] as well as international testing standards such as IEEE 829[6] or the terminology used by ISTQB[7]. The working group for model development used these sources and others in addition to their personal experiences in the field of Software Testing to come up with specific goals and practices for each process area that were then subjected to a thorough review by a broad review board.

Moreover, an Assessment Method Accreditation/Audit Framework for TMMi® as a Process Assessment Model (PAM) in accordance with ISO/IEC 15504 is being developed together with the process to certify commercial assessment methods against the standard model. This represents a crucial step in making the PRM applicable and any assessment results reusable and comprehensible. All resources can be downloaded free of charge from the TMMi Foundation homepage, www.tmmifoundation.org

## 3 TMMi Model Development

The working group members for Model Development and Maintenance come from a wide cross section of industry organizations and academic institutions. Working in virtual teams and a variety of input, together with a public review process comes with challenges, but also a much better opportunity to "get it right first" and facilitate an early adoption of the model through the inclusion of many feedback comments.

Since the concepts of TMM have already been published[4, 8], a major activity for the working group consisted of consolidating available sources and comparing them to existing industry and proprietary frameworks and process models to ensure TMMi's compatibility with these models. The working group set out to develop TMMi in 2 stages, focusing on describing the overall objectives of TMMi together with a detailed description of maturity levels 2 and 3 for a first release. Various supporting organizations and reviewers were invited, with the outcome being the release of TMMi level 2 in 2008 and level 3 in 2009. Using feedback gathered on these two levels and broadening the working group to enable more global participation, the full level 4 was then released in 2010 with the completion of level 5 planned for 2011. This continuous release schedule shows the commitment of the TMMi Foundation to consider a variety of inputs for the TMMi, and the commitment of each working group member and reviewer to stay involved in the model development and maintenance over a prolonged period of time. It should be noted that all this effort by the working group members and reviewers was given voluntarily and freely.

Each process area was defined by the working group at a higher level for a first review, consisting of the following elements:

- Purpose
- Introductory Notes
- Scope
- Specific Goal and Practice Summary

The working group undertook an extensive, cyclical peer review of the initial drafts before submitting to the independent review board. The review board consists of approximately 90 individuals who were signed up members of the TMMi Foundation without any specific membership requirements such as organizational alignment etc. The review board was provided with a feedback template and expectations for the review process, with the whole process being coordinated and facilitated by a working group member. Upon receiving input after a 4 week review period, each feedback comment was reviewed and either incorporated into the next draft or rejected by the working group. A second review cycle with the same review board then followed the definition of the specific and generic practices.

Particular focus was being put on not only fulfilling the requirements laid out by ISO/IEC 15504 regarding the fundamental elements of a Process Reference Model, but actually providing additional details for the purpose and expected outcomes of each process. The model not only looks for a pragmatic generic process and expected activities to be in place for an efficient and effective testing process but also looks to ensure it is fully deployed and used operationally within the organizational unit. This approach was taken with the objective of making TMMi readily usable to a wide audience, providing examples and references to other models. This approach can be seen in the following excerpt of a specific practice description for the Process Area PA3.4 Non-Functional Testing at TMMi level 3:

*SP 2.2   Define the non-functional test approach*
*The test approach is defined to mitigate the identified and prioritized non-functional product risks.*

*Typical work products*
*1.      Non-functional test approach (documented in a test plan)*
*The approach should be described in sufficient detail to support the identification of major test tasks and estimation of the time required to do each one.*
*Sub-practices*
*1.      Select the non-functional test techniques to be used*
*Examples of non-functional test techniques to be selected include the following:*
- *Heuristic evaluation, survey and questionnaires for usability*
- *Operational profiles for reliability*
- *Load, stress and volume testing for efficiency*

*Note that also black box, white box techniques and experienced-based techniques such as exploratory testing and checklists can be selected to test specific non-functional quality attributes.*

This attention to detail and broad input lead to some lessons learned as part of this journey. The commitment to timelines needs to be enforced by a central working group, especially when considering a large and dispersed review board coming from different backgrounds and viewpoints. In order to gain acceptance to the final work product each review comment needs to be analyzed and brought to closure, either by using the comment to update the actual model or by providing a traceable and comprehensible reason for rejection. Moreover, while the software testing industry is used to working in virtual teams, some time and expenses should be allocated to physical meetings by the core team in order to speed up the thought process and be able to foster a more creative working environment than one solely based on email and collaboration platforms. Finally, while parallel work streams seemed to be feasible for each process area (thus speeding up model development), the author had to realize that given the large interdependency of each process area within the model a sequential approach was more efficient and effective in the end.

## 4 Conclusions

Model development for a broad topic such as process improvement of testing and quality-related processes comes with a specific set of challenges. The iterative and collaborative way of working established by the TMMi Foundation proved to be an appropriate way of incorporating many different streams of thought while enabling the delivery of a high quality product. The TMMi Foundation is now working on finalizing the description of TMMi level 5 as well as the creation of a PAM as guidance for TMMi assessment processes.

**Acknowledgments.** The author would like to acknowledge the TMMi Foundation Working group for Model Development for all their dedication and commitment to delivering the TMMi level descriptions.

## References

1. Grottke, M.: Software Process Maturity Model Study. IST-1999-55017 (2002)
2. Schlich, M.: ASQF RFG Maturity Models Franken (2009), http://www.itprojectservice.de/Downloads/TMMI_20091006.pdf
3. IDC, Worldwide Discrete Testing Services 2009–2013 Forecast: Stepping Out from the Shadows, IDC #219959 (2009)
4. Burnstein, I.: Practical Software Testing, Springer Professional Computing (2002)
5. Gelperin, D., Hetzel, B.: The Growth of Software Testing. In: CACM, vol. 31(6), pp. 687–695 (1998)
6. IEEE 829, Standard for Software Test Documentation. IEEE Standards Board (1998)
7. van Veenendaal, E. (ed.): Standard Glossary. ISTQB (2010)
8. Burnstein, I., Suwanassart, T.: Developing a testing maturity model for software test process evaluation and improvement. In: Proceedings Int. Test Conference (1996)

# Experiences from Informal Test Process Assessments in Ireland – Top 10 Findings

Fran O'Hara

Inspire Quality Services, Dublin, Ireland
`fran.ohara@inspireqs.ie`

**Abstract.** Based on assessing and supporting improvements in the test processes of many organizations, the author has compiled a 'top 10' of recommendations that are frequently identified as 'quick wins'. These are practical suggestions that are typically low cost but high benefit in terms of solving problems with the testing process and helping to achieve improvement goals such as improved test effectiveness, reduced test execution time, etc. in support of higher level business goals such as quality improvement or productivity improvement.

## 1 Introduction

Given that testing typically accounts for 30-40% of the overall development effort, it is a key area to leverage in support of your business goals such as time or cost reduction or quality improvement. To give explicit support for this, test process improvement models have been developed to complement software/systems process assessment/improvement frameworks by providing additional detail on test related processes. In Ireland models such as TPI® and TMMi® have had some limited use in various industry domains but in general this use is limited to informal use (rather than organizations pursuing formal accreditation where available). Indeed the author has experienced many organizations that just look for expert independent assessments of their current practice with a view to providing practical recommendations to help solve their problems and achieve their goals. The use of a model is often not formally requested. However, in the authors experience of performing these informal assessments there are common patterns which are found and which have a strong correlation to the key areas in the models.

The main drivers for these assessments have included:

- A recent or current project that has/is not meeting expectations or commitments.
- High priority business/department drivers such as
- Cost reduction
- Improving Customer/Business/User Satisfaction
- Productivity increases
- Quality improvement
- The need to scale up operations and provide more structure
- Internal stakeholder dissatisfaction
- Poor management or operations visibility on project progress
- Inconsistency across projects
- A continuous improvement program

- A new hire or champion
- A new customer or project
- A customer demand or commitment
- Achievement of a defined standard

## 2 Experiences

An interesting observation is that a number of the key recommendations from the above test process assessments often relate to processes that interface/interact with testing. These broader issues and associated recommendations are addressed in systems/software process assessment/improvement frameworks but are often outside the scope of test specific models. This highlights a key learning point in not being too rigid when limiting the scope of an assessment to an area such as test. It also implies that the assessor(s) should ideally have some expertise in areas that bound the testing process rather than being test specific.

Based on assessing and supporting improvements in the test processes of many organizations, the author has compiled a 'top 10' of recommendations that are frequently identified as 'quick wins'. These are practical suggestions that are typically low cost but high benefit in terms of solving problems with the testing process and helping to achieve improvement goals such as improved test effectiveness, reduced test execution time, etc. in support of higher level business goals such as quality improvement or productivity improvement. Many of these principles and approaches are embedded in industry standards/models such as TPI® and TMMi®. Many of the recommendations also relate to elements of the ISTQB body of knowledge and to testing methodologies such as TMap®. It should be noted of course that the organizational context of course will result in often significant variations in the findings and recommendations and in particular the priorities will be specific to an organisation's business drivers and objectives.

This list is also influenced by more recent work where varying degrees of agile implementation are to be found which create variations on the findings/recommendations. This presentation will include these variations and draw conclusions for implications for the traditional test/software process improvement models which, at least on the surface, seem to relate to more plan-driven sequential lifecycles.

## 3 Top 10 Recommendations (Not in Any Order of Priority)

1. Improve the overall test strategy/approach
    o Clarify product risks associated with functional and non-functional requirements and then define the associated test levels needed to address them. This would include defining objectives and entry/exit criteria (including coverage measures) for each test level like unit test and system test to minimize duplication and maximize coverage.
    o For agile projects it is key to adapt the test strategy for iterative/incremental development (with greater need for regression testing and hence automation) and using approaches suitable for lightweight documentation

2. Use relevant test design techniques
3. Implement a risk-based approach to testing
    o A key element is the use of risk workshops with key stakeholders to identify and analyse the product risks to feed into testing
4. Improve vendor quality management in particular
    o Agree an integrated test strategy/approach
    o Agree review/milestone points and test related deliverables
5. Perform testing earlier in the lifecycle, in particular
    o Review requirements formally to identify major faults
    o Design test cases early as issues will be found in requirements as a side effect
    o Validate requirements with users (e.g. through validation prototypes)
6. Provide a lightweight definition of the testing process with supporting templates/tools and checklists
7. Improve test planning
    o Define and agree the key elements of a project plan for testing (see IEEE 829 for a template) to help ensure buy-in from relevant stakeholders and to get consensus on the balance between risk and cost/time/scope
    o Ensure integration with overall project plan
8. Improve test estimation
    o Actually part of planning above but often needs to be emphasized to consider for example using metrics based estimation using historical data and/or having adequate review of estimates
    o In an agile context needs to be included in overall team planning for example in Sprint planning in Scrum and incorporated into overall sprint velocity based on a definition of 'done' (effective exit criteria)
9. Ensure adequacy of test environments and configuration control thereof particularly for system test
10. Improve configuration management
    o This applies to configuration management of testware and of configuration items on the development side and ensuring integration and traceability between them.

# High Levels of Process Capability in CMMI and ISO/IEC 15504

Terry Rout

Software Quality Institute, Griffith University, Queensland, Australia
t.rout@griffith.edu.au

**Abstract.** The recent release of CMMI V1.3 incorporates a number of changes to the model and framework; one of the more interesting is the decision to do away with Capability Levels 4 and 5 in the Continuous Representation, while retaining the high levels of Organizational Maturity. This paper examines some of the issues that may have driven this decision, and explores the opportunity provided for greater interaction between CMMI and ISO/IEC 15504.

**Keywords:** ISO/IEC 15504, SPICE, CMMI, process capability, high maturity.

## 1 Introduction

The recent release of the V 1.3 of the CMMI [1] incorporates some interesting decisions on the structure and content of the family of process models. One of the more significant is the decision to do away with Capability Levels 4 and 5 in the Continuous Representation, while retaining the high levels of Organizational Maturity. At first sight, this appears to establish a degree of inconsistency between the two representations in CMMI; further, it seems to further separate the CMMI framework from that defined in ISO/IEC 15504. In this discussion, it is argued that none of these first impressions is correct; it will be shown that the decision in fact addresses an inconsistency existing in previous versions of CMMI, and that it offers a significant opportunity for greater interaction between CMMI and ISO/IEC 15504.

## 2 CMMI Model Structure

The principal elements of the CMMI models are the Process Areas; in V1.3, there are 22 Process Areas defined. A process Area is " a cluster of related practices in an area that, when implemented collectively, satisfies a set of goals considered important for making improvement in that area." [1, p 11]. The CMMI Product Team makes it clear that, while the CMMI models are based on achieving process improvement, and draw from a long tradition of theory and practice, the Process Areas are not intended to be equated to process descriptions – "CMMI models provide guidance to use when developing processes. CMMI models are not processes or process descriptions." [1, p5].

CMMI models have to "representations"; a Staged Representation, with the Process Areas clustered to define Maturity Levels (from 2 to 5), and a Continuous Representation in which each Process Area is viewed as achieving a Capability Level.

In CMMI models up to Version 1.2 [2], there are 5 Capability Levels; in Version 1.3, however, the scale of Capability Levels extends only to CL3. This decision resulted from one of the key issues driving the revision – to improve high maturity material in the models.

The fact that a Process Area is not, and is not intended to be, a process, sets up an issue of consistency in addressing the assessment of higher levels of process capability – CL4 and CL5 – in V1.2 and earlier models. A formal mapping of CMMI (V1.1) to the processes defined in ISO/IEC 12207 [2] demonstrates that Process Areas map to multiple processes, and the same process can be identified in multiple Process Areas. Because of the high degree of granularity of most CMMI Process Areas, it is difficult to establish empirically that the entire Process Area is "quantitatively managed". To rate a complete Process Area as achieving CL4 or CL5 would require that quantitative management had been achieved in a majority of the processes addressed in the Process Area – a difficult (if not impossible) factor to demonstrate. This means that it is quite possible for an organization to achieve Maturity Level 4 (or 5) without any of its Process Areas being assessed at Capability Levels 4 or 5.

In the establishment of high maturity, this issue is addressed satisfactorily, by limiting the required scope of quantitative management to "sub-processes", elements of Process Areas that are more similar to the life cycle processes of ISO/IEC 12207 or ISO/IEC 15288. Prior to the release of V1.3, it was quite possible for an organization to achieve Maturity Level 4 (or 5) without any of its Process Areas being assessed at Capability Levels 4 or 5. With the release of CMMI V1.3, this issue has been addressed by removing Capability Levels 4 and 5 from the scale in the Continuous Representation. The requirement to establish that quantitative management of "selected sub-processes" is established is still present, however, and applicable to the establishment of Maturity Levels 4 and 5. With the restriction to the Capability Scale, the previous inconsistency has been removed; however, it leaves unaddressed the issue of how the achievement of quantitative process management can be best demonstrated.

## 3 Process Capability in ISO/IEC 15504

ISO/IEC 15504-2 [4] defines a Measurement Framework for process capability, defined as "a characterization of the ability of a process to meet current or projected business goals". The framework provides a six point ordinal scale for assessment of capability; detailed empirical studies through the SPICE Trials have demonstrated the internal consistency and predictive validity of the scale [5, 6].

The measurement framework for Capability Levels 4 and 5 are strongly aligned to the Generic Practices for CL4 and CL5 in V1.2 of CMMI [4], and describes the characteristics of a quantitatively managed and quantitatively improving process. The processes to which the framework are applicable cover a wide range of functions in the Information Technology domain, and have been shown to be applicable in other domains. Process models are currently available that address software engineering (ISO/IEC 15504-5), systems engineering (ISO/IEC 15504-6) and the automotive domain; models under development will address IT service management (ISO/IEC 15504-8), safety engineering (ISO/IEC 15504-10) software testing (ISO/IEC 33063) and medical device software.

The processes (in an ISO/IEC 15504 assessment) are well defined entities, described in terms of the purpose of performing the process, and the identified outcomes of performance. The granularity of the models is in general such that it is not difficult to equate these well-described processes to the "selected processes or sub-processes" specified in the CMMI High Maturity Process Areas. The assessment of process capability in selected processes as CL4 or CL5, then, can be seen as confirming achievement of key aspects of these High Maturity Process Areas.

## 4   Conclusions

The revision of the high maturity features – and specifically the removal of Capability Levels 4 and 5 from the Continuous Representation of CMMI – can be seen to provide an opportunity for merging aspects of CMMI appraisals and 15504 assessments, in order to provide a higher standard of proof of achievement of high levels of maturity. The demonstration of Capability Levels 4 and 5 specified in ISO/IEC 15504 clearly requires the same or similar performance to the establishment of "quantitative management of sub-processes", required in CMMI appraisals.

It would seem logical, then, that a combination assessment – incorporating ISO/IEC 15504 assessment of selected individual processes, whether from software (Part 5), system (Part 6) or service management (Part 8) processes, would provide ample evidence for the establishment of the degree of quantitative management expected in a CMMI appraisal. In particular, the application of conformity assessment in the domain of 15504 assessments of process capability [7] will enable solid, certified evidence of achievement of quantitative management of process performance and improvement to reinforce the evidence of satisfaction of the CMMI high maturity process areas.

## References

1. CMMI Product Team, CMMI for Development, Version 1.3. SEI Technical Report CMU/SEI-2010-TR-033 (November 2010)
2. CMMI Product Team, CMMI for Development. Version 1.2. SEI Technical Report CMU/SEI-2006-TR-008 (August 2006)
3. Rout, T.P., Tuffley, A.: Harmonizing ISO/IEC 15504 and CMMI, Softw. Process Improve. Pract. 12, 361–371 (2007)
4. ISO/IEC 15504-2, 2003. Information technology – Process assessment – Performing an Assessment
5. El-Emam, K.: The Internal consistency of the ISO/IEC 15504 software process capability scale. In: Proceedings of the 5th International Symposium on Software Metrics, pp. 72–81 (1998)
6. Rout, T.P., El Emam, K., Fusani, M., Goldenson, D., Jung, H.-W.: SPICE in retrospect: Developing a standard for process assessment. Journal of Systems and Software 80, 1483–1493 (2007)
7. Rout, T.P.: The evolving picture of standardisation and certification for process assessment. In: Proceedings of QUATIC 2010, Conference, Porto, Portugal (2010)

# Process Innovation Reaping Customer Satisfaction

B. Sridhar and G. Rajesh

Embedded Systems SBU, CMC Ltd, India
{B.Sridhar,Rajesh.G}@cmcltd.com

**Abstract.** CMC, part of the TATA Group, is a leading Embedded, Engineering and IT consulting firm based in Hyderabad, India. CMC is customer focused with an emphasis on quality which is evidenced through the achievement of ISO 9001, SEI-CMMI Level 5 and Auto SPICE Level 3 certifications for its operations in all of its delivery centers. CMC has been continuously striving for Process innovation with the ultimate goal of achieving productivity improvements to the delight of customer. CMC observes that process innovation has a direct correlation to the customer satisfaction levels. Coupled with the effective project management, vision driven leadership has shown an increasing trend in the customer satisfaction index. As a matured organization it was imperative that a sustainable rewarding culture for process improvements was developed and nurtured. Process improvements at CMC are triggered with the objective of improving granular level planning and in order to improve a process we change the tasks. Tasks may be eliminated or combined. The sequence in which they are performed may be changed. The location where they are performed or the people doing them may be changed. And, the method of accomplishing them may be changed, often by changing tools and equipment. When these changes are well conceived they can produce positive results in two ways, - better results and lower costs.

## 1 Introduction

The Automotive industry predominantly deals with around 80% of code/ components being reused. With the recent bigger turn down in the automotive industry, the challenge and onus is laid on cost effective product delivery. While testing contributes around 30% of the complete development lifecycle, there is tremendous dependency on its outcome for the overall product delivered by the customer to its clients. Therefore, there is great onus and responsibility on CMC not only to complete the customer requirements but also to meet the industry demanding cost and quality objectives. This is achievable amongst other things through Process innovation.

## 2 Background, Approach , Process Innovations

Leads across the projects are handling tasks with the focus of process innovation and teams have initiated many improvement steps that resulted in Improvement in quality of deliverables, cycle time reduction and reduction of efforts needed to execute the projects. Proactive risk identification enables identifying improvements opportunities.

Most often the risk management process plays a pivotal role in enabling improvements, the project teams at the inception of the project are tasked with the job of identifying potential risks (identified by brainstorming and case scenario ways). During the process of identifying mitigation plans, the project teams envisage the option of translating the mitigation plan into an improvement initiative.

Additionally, a detailed analysis of customer feedback (collected from direct and indirect means from about 100¬¬+ clients) was performed and this has revealed that the customer delight was primarily because of cycle time reduction and zero defect delivery. Also the root cause analysis (RCAs) performed over low CSI identified the fact that rework and schedule slippage have been the major contributors. As the automotive electronics group is certified for AutoSPICE, periodic SPICE assessments and SPICE framework training programs have the provided the technology teams with a new dimension of improving the engineering processes.

We provide below a few cases that demonstrate the productivity improvements / Process innovations.

- As part of the software development process, there was a huge dependency on the hardware availability for the testing activity to begin, sometimes resulting in schedule slippages. In consultation with the customer, CMC has introduced/ implemented simulation-based development and testing where the hardware peripherals are simulated and this simulation has been used for the testing of ECU software. In this process, most of the software development/ testing can be completed; errors in the software can be detected and fixed before the availability of hardware. This has reduced on-target-testing, debugging time and most importantly rework effort, thereby saving time and cost.
- One of the activities performed under Braking projects is the communications I/O functional testing. This requires a detailed test plan to be prepared from an input file called CAN dbc. The dbc file typically comprises of large number of functional messages used by various nodes in the in-vehicle network. Earlier, this activity of test plan creation was manually done by a tester by reading each signal from the dbc file and copying the same in to the test plan workbook. This activity consumed few days of laborious work and was prone to lot of errors. In order to overcome this problem, CMC proactively came up with an innovative idea of developing a tool that would simply parse the entire dbc format file and populate all the signal requirements in the test plan in few seconds. This idea was very much appreciated by the client and lead to improvement in quality and productivity.
- Air-bag Verification & Validation projects need crash signals as inputs to validate the performance of the Electronic Control Unit (ECU). These signals were earlier generated using the arbitrary waveform generators (AWG). These generators were limited in number and had to be shared among multiple projects resulting in delivery schedule pressures. CMC teams have proactively developed special hardware boards customized for these projects. These are relatively less expensive and can be locally fabricated.

Productivity (no of test cases executed/day) Improvement in each category.

The availability of these boards resulted in quicker execution of the projects with less pressure on the teams as the wait-time for the AWGs has been avoided.

Hardware spy tool is an example of an in-house tool developed by CMC engineers, who have helped in effective usage/ sharing of the CANALYSER/CAN Cards among the teams. Three tier architecture was used for developing this application.

## 3 Architecture of the SPY Tool

The application consists of the following package.

- Window services developed in VB.NET, which shall extract the details of hardware resource like serial number and PC's Mac address.
- Centralized database server using SQL SERVER 2000 to store hardware's and user's data which always gets updated through the windows services installed in all PCs'.
- Web Application which will display the information of all the available resources, by extracting data from the centralized database server.

This Innovation helps to

- Effectively track and utilize the available 'CAN' Hardware resources across windows network.
- Helps in effectively Plan and execute the projects where hardware resources like digital meters, CAN hardware, Barcode readers, Printers etc. are shared by multiple users.
- Prevents manual tracking of resources there by saving lot of productive time of both the employees and managers
- This has resulted in significant cost reduction (approximately USD 0.6 million in 2 years time for a specific customer and this can be deployed across projects

As part of Design for Manufacturability guide lines process every design need to under go the Conductive Anodic Filament (CAF) violation check. The previous process for CAF violation check script was taking more time for execution and manual interaction was required. CMC has proactively developed new script which has reduced the manual work and reduced the execution time almost 150 times. The previous CAF violation script was executed on exporting the data using report writer which is in Microsoft excel (xls) file. The new script is executed on PCB design application and used the application memory that improved the performance and reduced the manual work of exporting the data in to Microsoft excel. This has been well acknowledged and appreciated by the customer.

The tools used for independent verification and validation did not provide entire set of scripts for code verification to verify compliance for software implementation guidelines. The guideline documents contained the rules derived from experience, where the tools did not really help. So it involved manual verification and more effort. CMC took up the task of updating the scripts database so that this entire manual effort is eliminated. Today we have database that can cater various coding rules besides the Misra.

In challenging scenarios, where time to market is one of the key aspects of delivery management, the Project Managers had the challenge of optimal utilization of resources, this was augmented with the help of burn down charts, through understanding of the work loads and effectively managing them. Routine health checks and reviews have also triggered process improvements in terms of updates to the checklist and training materials.

One of the key drivers for achieving the process improvements and innovations has been a strong training program – technology coupled with process training with the basic awareness in the delivery team about customer landscape and its drive towards cost/ quality delivery. The HUB SPOKE model approach is being implemented for building the competency levels of the team,

- Core team in the Automotive Electronics group across the verticals Hardware, Software Verification and validation having expected expertise mobilized internally
- Each member of core team assigned as a coach and mentor to 4-5 member technical team
- Ramp-up in any of the verticals is done by adding a new Hub-Spoke Unit (HSU) each time.
- Mobilization done through lateral hires, up-skilling of internal associates, academic institutes and leveraging alliances.

# 4 Conclusions

It has been observed that process innovation has a direct correlation to the customer satisfaction levels. Coupled with the effective project management, vision driven leadership has shown an increasing trend in the customer satisfaction index. Repeat business and an increase in the employee motivation levels (resulted from internal employee satisfaction survey) have been observed. Organizations need to focus on process improvements as these lead to better motivation and lead up to the challenges for employees to better their performance, thereby leading to an increased customer satisfaction.

# Process Improvement in an R&D&I Center Using *Enterprise SPICE* and *SPICE for Research* Models

Clênio F. Salviano

CTI: Centro de Tecnologia da Informação Renato Archer
Rodovia D. Pedro I, km 143.6, CEP 13069-90
Campinas, SP, Brazil
`Clenio.Salviano@{cti.gov.br,gmail.com}`

**Abstract.** This article presents objectives, strategy and early results of a process improvement experience using Enterprise SPICE and SPICE for Research models in a 12-months process improvement cycle (from January to December 2011). This cycle has been performed in a Research, Development and Innovation (R&D&I) Division on Software Quality and Process Improvement of CTI Renato Archer, a Brazilian Information Technology R&D&I Center. This process improvement cycle is using values and principles of process improvement and PRO2PI Methodology (Process Modeling Profile to drive Process Improvement) to guide the process improvement.

**Keywords:** Process Improvement, SPI Manifesto, PRO2PI Methodology.

## 1 Introduction

DMPQS is a Research, Development and Innovation (R&D&I) Division on Software Quality and Process Improvement of *CTI Renato Archer*, a Brazilian Information Technology R&D&I Center (www.cti.gov.br). In 2009, DMPQS R&D&I activities included the usage of Systematic Literature Review technique and the production of Technical Reports, the participation in a Digital Convergence Institute. The activities also included the decision to converge its current efforts towards a Digital Convergence focus as part of CTI 2011-2015 strategic planning. Following its continuous improvement, DMPQS is been performing a process improvement cycle from January to December 2011.

This process improvement cycle is using the three values and ten principles of Software Process Improvement defined in the SPI Manifesto [1] and the PRO2PI Methodology (**Pro**cess Modeling **Pro**file *to* drive **P**rocess **I**mprovement) [2]to guide the process improvement. Following PRO2PI method for process improvement cycle (PRO2PI-CYCLE), one of the first actions was to define the organizational objectives for the cycle. The main objective is related with improving the current improvement efforts towards more integration with the upper levels strategic objectives and towards better results on integration R&D&I. The strategy was to act in both a strategic vision and more operational actions in order to include all integrants of DMPQS.

## 2 Process Capability Profile to Guide Process Improvement

One of the first actionsis to establish a Process Modeling Profile to guide the process improvement. This profile is composed of three types of models: Process Capability Profile, Process Enactment Description and Process Performance Indicator. The term "establish" is used in PRO2PI with a specific and broad meaning, including: identification of organizational context and objectives, identification of relevant reference models, definition of a Process Modeling Profile with specification models for the improved process, an analysis of the current processes in terms of the specification models and a revision of the specification models. These models cover capability profile, enactment descriptions and performance indicators.

This section describes the establishment of Process Capability Profile. For Process Capability Profile specification model the decision was to use the Enterprise SPICE model [3] (Figure 1)as the main reference. Enterprise SPICE® is an integrated model for enterprise-wide assessment and improvement for use with international standard ISO/IEC 15504 (SPICE). Therefore Enterprise SPICE is a SPICE PAM (SPICE compatible Process Assessment Model). It provides an efficient and effective mechanism for assessing and improving processes deployed across an enterprise. Enterprise SPICE defines 29 processes organized into 4 categories (Figure 1).

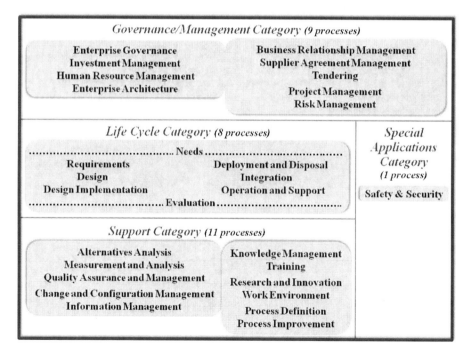

**Fig. 1.** Enterprise SPICE´s Categories and Processes

From an analyses of the objectives and the current situation, two Enterprise SPICE processes (Enterprise Governance and Knowledge Management) were selected, both at capability level 2, as the Process Capability Profile specification model for this cycle. An Enterprise SPICE process, as any other SPICE PAM process, is actually a reference process to guide the establishment of processes in the organization to achieve the reference process purpose.

The purpose of the Enterprise Governance process is to guide the establishment of processes to "establish strategic enterprise direction and ensure the enterprise achieves its goals and objectives".

The purpose of the Knowledge Management process is to guide the establishment of processes to "ensure that individual knowledge, information and skills are collected, shared, reused and improved throughout the organization".

Another decision was to use the *SPICE for Research* model[4] as a complementary reference because, for DMPQS´s context (R&D&I Division) it defines lower level processes to help the implementation of some aspects of the selected Enterprise SPICE processes. SPICE for Research is a SPICE PAM developed by CTI and Unicamp (State University of Campinas, www.unicamp.br) for University Research Laboratory (URLab). An URLab is a unique environment that performs knowledge-intensive activities. It needs systematic organization in its management processes to consider a satisfactory integrated vision associating the strategy, mission, people, culture, infrastructure, and mainly knowledge actions. SPICE for Research defines 25 processes organized into 6 process groups (Figure 2).

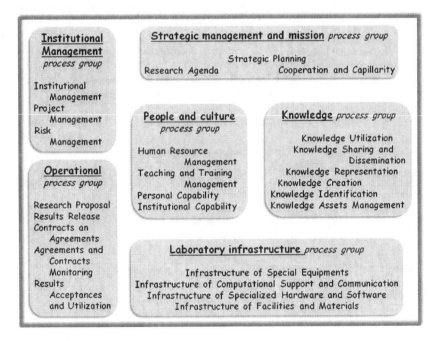

**Fig. 2.** SPICE for Research´s Process Groups and Processes

For Enterprise SPICE´s Enterprise Governance process, all three processes from SPICE for Research´s Strategic and Mission process group were selected

(*StrategicPlanning*, *Research Agenda* and *Cooperation and Capillarity*), all three at capability level 2. The purpose of Strategic Planning process is to guide the establishment of processes to "Establish a common reference for the vision, mission and strategic goals in conformance with higher level strategic plan such as the faculty's and university's plans [and the R&D&I Center strategic planning]". The purpose of Research Agenda process is to guide the establishment of processes to "Establish a common reference for the research agenda in conformance with the URLab[or R&D&I Division] strategic planning and its competencies". The purpose of Cooperation and Capillarity process is to guide the establishment of processes to "Identify, establish, coordinate and monitor the cooperation, formal and informal relationships and also the capacity toinfluence in other entities and external agents with ethicsand legal limitations".

For Enterprise SPICE´s Knowledge Management process, two processes from SPICE for Research´s Knowledge process group were selected:*Knowledge Identification*(at capability level 2) and *Knowledge Representation*(at capability level 3). The purpose of Knowledge Identification process is to guide the establishment of processes to "Guarantee the identification of existing knowledge or the necessary for future actions of the strategic goals". One action is to establish Systematic Literature Review as a basis for external knowledge identification for all major R&D&I efforts. The purpose of KnowledgeRepresentation process is to guide the establishment of processes to "Stimulate and guarantee that the knowledgegenerated is represented to be useful for others". One action is to establish a standard process to represent all relevant R&D&I results as Technical Reports.

## 3 Conclusions

A SPICE PAM, such us Enterprise SPICE and SPICE for Research, is target to be used as reference for process assessment. However, it can be, and it has been, used also as a reference for improvement. There is a third perspective to complete the usage of a SPICE model: be a reference to think about an organization. This third perspective has been already performed with success in this process improvement cycle. At the end of 2011, the cycle is planned to be completed and a formal assessment is planned for 2012.

## References

1. Pries-Heje, J., Johansen, J. (chief eds.): SPI Manifesto, eurospi.net, version A.1.2 (2010)
2. Salviano, C.F.: A Multi-Model Process Improvement Methodology Driven by Capability Profiles. In: Proc. of IEEE COMPSAC, Seattle, USA, pp. 636–637 (2009), doi:10.1109/COMPSAC.2009.94
3. The Enterprise SPICE Project Team. Enterprise SPICE (ISO/IEC 15504) An Integrated Model for Enterprise-wide Improvement. Technical Report – Issue 1 (September 2010)
4. Silva, J.V.L., Nabuco, O.F., Salviano, C.F., Reis, M.C., MacielFilho, R.: Towards an ISO/IEC 15504-based Process Capability Model for Public University's Research Laboratory. In: Proc. Fourth Int. SPICE Conference, South Korea, pp. 12–21 (2007)

# Med-Trace

Fergal McCaffery and Valentine Casey

Regulated Software Research Group,
Dundalk Institute of Technology & Lero, Dundalk, Co Louth, Ireland
{fergal.mccaffery,val.casey}@dkit.ie

**Abstract.** Traceability is central to medical device software development and essential for regulatory approval. To achieve compliance an effective traceability process needs to be in place. This is difficult to achieve due to the lack of specific guidance which the medical device regulations and standards provides. This has resulted in many medical device companies employing inefficient software traceability processes. In this paper we briefly outline the development and implementation of Med-Trace a lightweight software traceability process assessment and improvement method for the medical device industry.

**Keywords:** Medical Device Software Traceability, Software Process Improvement, SPI, Lightweight Assessment Method, SPICE, CMMI®.

## 1 Introduction

The important role that software plays in medical devices continues to increase. This has taken place in conjunction with the demand for increased medical device functionality. As a result of both of these factors the complexity of medical device software and its development has also increased [1]. This has necessitated the requirement for effective traceability and risk management processes and tools to be in place to facilitate the development of medical device software.

Medical device companies must ensure that they comply with medical device regulations as governed by the region in which they wish to market their device. If a device is to be marketed in Europe the medical device should be developed using processes that comply with the European Council's Medical Device Directive (MDD) (1993/42/EEC) [2] and amendment MDD (2007/47/EC) [3]. When a medical device is to be marketed in the United States (US) the medical device should be developed using processes that comply with the Food and Drug Administration (FDA) guidelines. In both locations the medical device companies must be able to produce sufficient evidence to support their product's compliance.

In addition to achieve compliance national regulatory requirements also recommend conformance to a number of international standards which include: IEC 62304:2006 [4], ISO 14971:2007 [5], ISO 13485:2003 [6] and IEC 62366:2007 [7]. Given the need to address the requirements of national regulations and international standards software medical device companies are focused on compliance. While this is essential to market their products it has resulted in a lack of emphasis on process improvement and the achievements of its associated benefits [8].

## 2 Software Traceability

Software traceability refers to the ability to describe and follow the life of a requirement in both a forward and backward direction. This includes from its origins, specification, development, subsequent deployment and use and through periods of on-going refinement and iteration in any of these phases. The deployment of an effective traceability process is essential to facilitate the development of high quality software systems [9]. Therefore software traceability is central to medical device software development and essential for regulatory compliance.

In order to comply with the regulatory requirements of the medical device industry it is necessary to have clear linkages and traceability from requirements - including risks and hazards - through the different stages of the software development and maintenance lifecycles. The regulatory bodies request that medical device software development organizations clearly demonstrate how they follow a software development lifecycle without mandating a particular lifecycle. This is further compounded by the requirement to adhere to numerous standards without guidance on how they should be implemented. Given the lack of guidance and importance that traceability plays in medical device software development it was recognized that this was an important area which needed to be addressed. The authors decided to tackle this issue by developing a lightweight assessment method called Med-Trace, specifically to assist companies to adhere to the traceability aspects of the medical device software standards and regulations and also to improve their process.

## 3 Med-Trace and Observations from Two Assessments

Based on the results from an extensive literature review, the relevant areas of the CMMI® [10], ISO/IEC 15504-5:2006 [11] and previous experience of developing lightweight process assessment methods Med-Trace has been developed. Med-Trace is a lightweight assessment method that provides a means of assessing the capability of an organization in relation to medical device software traceability. It enables software development organizations to gain an understanding of the fundamental traceability best practices based on the software engineering traceability literature, software process models, and the relevant medical device regulations and standards. Med-Trace may be used to diagnose an organization's strengths and weaknesses in relation to their medical device software development traceability practices. The goal of a Med-Trace assessment is not certification, but to assist medical device organizations to improve their software development traceability process.

The Med-Trace assessment method contains eight specific stages. The assessment team normally consists of two assessors who share responsibility for conducting the assessment. **Stage 1**, a preliminary meeting between the assessment team and the company wishing to undergo a Med-Trace assessment takes place. During **stage 2**, the lead assessor provides an overview of the Med-Trace assessment to members of the organization. At **stage 3** a review is undertaken of project documentation. Staff from the organization with responsibility for traceability are interviewed at **stage 4**. At **stage 5** the assessors jointly develop the findings report. **Stage 6** involves presenting the findings report. **Stage 7** is the collaborative development of a pathway towards achieving highly effective and regulatory compliant traceability practices. Having participated in the development of this pathway the organization are responsible for its implementation. **Stage 8** involves revisiting and reassessing the company approximately 3 months after the completion of stage 7 and reviewing progress against the recommended improvement path, a final report is also produced.

Two Med-Trace assessments have taken place. The first was in an Irish medical device organization, Medical Electronic (a pseudonym). Medical Electronic develop electronic based medical devices that are marketed in the US and Europe. The company recognized the importance traceability plays in medical device software development and they sought a lightweight assessment method to obtain guidance as to how they could improve their traceability process. The second assessment took place in North Medical UK (a pseudonym). The company develop electronic-based medical devices that require compliance with both the FDA and the MDD. North Medical UK also sought a resource-light assessment method to obtain guidance as to how they could improve their software development traceability process.

As a result of both assessments it was clear the organizations recognized the importance traceability plays in medical device software development. This was reflected in the fact that in each, a member of the management team was responsible for its implementation. The lack of detailed guidance on how to implement traceability was highlighted by the management of Medical Electronic and North Medical UK. While these organizations both employed a process for traceability, in each case these needed to be improved and formalized. The requirement for relevant training and the ability to record and leverage best practice with regard to traceability emerged. The serious limitations of utilizing manual tools such as MS Office to manage traceability was also recognized and needed to be addressed by both organizations.

The findings from the assessments identified important areas where improvements were required and these were confirmed in consultation with the management and staff of both organizations. The adoption of the development pathway provided realistic goals and the collaborative process provided motivation for their achievement. Both organizations are implementing their respective development pathways and have agreed to be reassessed as part of stage 8 of the Med-Trace assessment method.

## 4 Conclusions

In this paper we have presented Med-Trace a resource light process assessment method for the medical device software industry that can pinpoint specific areas for improvement with regard to traceability. We will continue to refine Med-Trace based on the experience gained in undertaking future assessments, interaction with medical device software organizations and discussions with medical device regulatory bodies. It is envisaged that further research will be undertaken for the development of similar lightweight software process assessment methods in the future.

**Acknowledgments.** This research is supported by the Science Foundation Ireland (SFI) Stokes Lectureship Programme, grant number 07/SK/I1299, the SFI Principal Investigator Programme, grant number 08/IN.1/I2030 (the funding of this project was awarded by Science Foundation Ireland under a co-funding initiative by the Irish Government and European Regional Development Fund), and supported in part by Lero - the Irish Software Engineering Research Centre (http://www.lero.ie) grant 03/CE2/I303_1.

## References

1. Rakitin, R.: Coping with defective software in medical devices. Computer 39(4), 40–45 (2006)
2. European Council, Council Directive 93/42/EEC Concerning Medical Devices. Official Journal of The European Communities, Luxembourg (1993)
3. European Council, Council Directive 2007/47/EC (Amendment). Official Journal of The European Union, Luxembourg (2007)
4. IEC 62304:2006, Medical device software—Software life cycle processes. IEC, Geneva, Switzerland (2006)
5. ISO 14971:2007, Medical Devices — Application of risk management to medical devices. ISO, Geneva (2007)
6. ISO 13485:2003, Medical devices — Quality management systems — Requirements for regulatory purposes. ISO, Geneva, Switzerland (2003)
7. IEC 62366:2007, Medical devices - Application of usability engineering to medical devices. IEC, Geneva, Switzerland (2007)
8. Mc Caffery, F., Casey, V.: Med-Adept: A Lightweight Assessment Method for the Irish Medical Device Software Industry. In: Software Process Improvement - European Systems & Software Process Improvement and Innovation Conference (EuroSPI), Grenoble, France (2010)
9. Espinoza, A., Garbajosa, J.: A Proposal for Defining a Set of Basic Items for Project-Specific Traceability Methodologies. In: 32nd Annual IEEE Software Engineering Workshop, Kassandra, Greece (2008)
10. CMMI Product Team, Capability Maturity Model® Integration for Development Version 1.2, Software Engineering Institute (2006)
11. ISO/IEC 15504-5:2006, Information technology — Process Assessment — Part 5: An Exemplar Process Assessment Model. ISO, Geneva, Switzerland (2006)

# Functional Safety – SPICE for Professionals?

Bernhard Sechser

Method Park Software AG, Erlangen, Germany
Bernhard.Sechser@methodpark.com

**Abstract.** The paper describes a possible way how to extend an existing process to fulfill the demands of Functional Safety from a process point of view. A systematic proceeding shall assure that the system under development reaches a sufficient small residual risk before it goes on to series production. Based on the existing company process at Continental several aspects will be scrutinized. In the end you will have a good overview of the similarities and differences of SPICE and Functional Safety and a first idea how to improve your processes to proceed on a safe way.

**Keywords:** SPICE, Functional Safety, ISO FDIS 26262, ISO/IEC 15504-10.

## 1 Introduction

"SPICE is dead – long live Functional Safety!" Do you know such slogans? Especially – but not only – in the automotive industry many people think like that because of the upcoming ISO 26262 standards. But does this really mean that SPICE needs to be used no longer? Of course: NO!

The problem rather is that many companies are already happy to demonstrate a SPICE level 2 to their customers. That result can be easily achieved because it needs only a single project that proceeds according to the demands of SPICE level 2. But starting with level 3, the organization itself enters the game, and here the above described projects will fail. The introduction of Functional Safety addresses an organizational top-down approach right from the beginning. Companies such as Continental have recognized this early and responded accordingly. You will see that, based on a lived company safety culture, the re-use and improvement of SPICE-proven processes are important steps towards this goal.

## 2 Background

Functional Safety is a strong requirement that has to be fulfilled if you want to release new products that can cause harm to people and environment. You have to evaluate and reduce the risk of a product before you put it on the market.

Functional Safety does not only cover the software part of a system but also the electronics part. Yet in contrast to software components hardware parts have a specific "feature": they age. In the course of time the probability that a hardware part fails becomes higher and higher. Software cannot age. Without doing modifications software remains in the same state forever.

So what could be a reason that unmodified software fails? Right: the faults are there from the beginning. The task is to eliminate these faults, or better not to make any faults. And what is the best way to assure this? A mature and practicable process – doesn't this sound familiar? If you want to check whether your process is able to produce correct software or not, you can use a process reference model, combine it with a capability scale and use both to perform an assessment. And here we are again: SPICE!

Thus we see that the evaluation of its own processes against SPICE is a good basis for safety-related development. However, it still needs some enhancements in order to cover all process relevant aspects.

## 3 Project Management Starts with a Safety Culture

Project management is one of the most important processes in SPICE as well as in Functional Safety standards. You need it to plan, control, adjust, and report your project related activities. In addition, the ISO/FDIS 26262 demands, among other things, that the project independent "overall safety management" guides the project teams through every single step they have to take within their development work. Within the organization awareness shall be established that safety is not just the concern of a single project but of the whole organization. A clearly defined strategy is as important as an organization-wide training program to ensure the competence of project staff. The requirement for independent audits by persons outside the project or in worst case even outside the organization forces to abandon the project-specific way of thinking. Of course, the safety standards refer primarily to the security-related activities and roles. But is it therefore reasonable to treat all other activities with less attention? No, of course not.

Continental has already started to establish an appropriate safety culture before the first draft of the ISO 26262 was available. A company-wide Safety Manager makes sure that the internal processes will conform to current safety standards. He ensures that the company's management is always aware of the importance of this issue. He takes care that product risks that are taken into account due to cost reduction measures must not be allowed. He communicates with independent certification bodies as well as with business-specific safety managers. Company wide training programs and defined role descriptions are passed to the business unit safety responsibles for the further implementation in their area. The Business Unit Safety Manager is the interface to the project teams. He offers them support in the implementation of the safety requirements, e.g. in the preparation of the project safety plan or the selection of the appropriate tool chain. He also takes care of the feedback from the projects by gathering and analysing the usability of the defined processes or the failure rates of the developed products. If all this was done properly, the Project Safety Manager - the classical Project Manager from SPICE – has to plan only "some" additional Safety Activities.

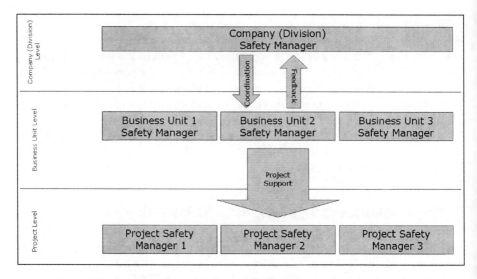

**Fig. 1.** Ensure a "Culture of Safe Working" in your organization

## 4 I Know What to Do – But How?

To work out the Safety Integrity Level (SIL) of a product respectively to decompose and assign it to the parts of a system by performing a hazard and risk analysis is a very important extension to the activities you have to plan and perform in a project. Only with the results from these analyses you are able to define the exact activities you need to do in your project. But not enough: by listing appropriate methods safety standards also give you a recommendation how to reach these goals. Is it sufficient, for example, to perform work product reviews by using a less formal walkthrough? Or do you have to perform a more formal method such as an inspection? Who should perform the review at all? A colleague from your own team, or a developer from the next project, or even a person from another organizational unit? SPICE says nothing about that. SPICE demands to review work product and to plan them in the project, but it leaves open how and by what method.

At Continental it was necessary to expand the existing process meta model by another component: the methods. Now they are able to make a selection from the existing method descriptions along with a SIL-dependent tailoring system.

## 5 Never Change a Safe System!

Based on the hazard and risk analysis a specific process safety life cycle including the SIL-related methods was selected. Any change after start of development can cause a repetition of the whole effort. Therefore – particularly in safety-related systems – it is very important to specify most of the requirements at the beginning of the project and to mark the safety-related parts with dedicated attributes. Any modification can increase the likelihood that an error creeps in.

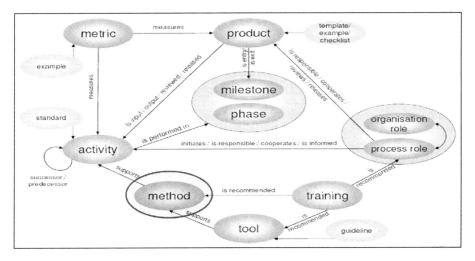

**Fig. 2.** Necessary meta model extensions for method selections

Experienced project managers at Continental know of course that customers like to introduce changes to the software until just before start of series. These last-minute changes have a direct influence on the costs of development. Modification costs will incur only once, and updates can be flashed in an easy way. Hardware changes in the late phases will be carried out only in emergencies. Only if you stop the change management process early, you will have the chance to generate a stable and error-free software version. Since unmodified software is not subject to random failures in the course of its use, you can focus on the process and reduce the risk of systematic errors.

Hardware components, however, are subject to natural aging. With its use, the risk is growing that a component fails dangerously. ISO 15504 is excluding the topic of hardware development, so far. So you would expect that the most innovations are probably in this area. However, the comparison with the already existing rules at Continental has shown that just in this domain a lot of analyses are already performed to determine failure rates and component behaviour.

## 6 A Release Is Just a Release?

However, by the end of a development cycle at the latest you should ask yourself whether the product your team has developed and is about to be shipped to the customer is really doing what it should do. A release is issued quickly. Errors that occur at the customer are annoying and can cost a lot of money. But with money you cannot replace all damages. Products can endanger people's life, whether by a no longer functioning vehicle brake or wrong calibrated radiation device at the hospital. If you are in doubt that there really has been done everything that is necessary to ensure safety, you should forbear to deliver the product and continue minimizing the risk by further test and analysis activities.

## 7 Conclusions

Respecting the above mentioned aspects it will not be very hard to use a SPICE conformant process to develop safety relevant products. However, we must not forget that we have dealt mainly with the process aspects. Functional Safety, of course, refers to a much larger part to the product itself. SIL-dependent selection of architectures and components should be addressed as well as the process aspects. Part 10 of ISO 15504, which is under development at the moment, contains some important additional processes on Functional Safety, which are helpful for the process capability determination as part of a SPICE assessment. However, it is necessary to always include a technical professional to determine and confirm the completeness of all safety goals.

First combined SPICE and safety assessments at Continental have shown that both aspects can complement each other usefully. However, in the end the product has to prove its safety.

**Acknowledgments.** Thanks to all people at Continental supporting me in evaluating and improving Functional Safety aspects in safety-relevant projects.

## References

1. Höhn, H., Sechser, B., Dussa-Zieger, K., Messnarz, R., Hindel, B.: Software Engineering nach Automotive SPICE. dpunkt.verlag (2009)
2. Sechser, B.: The Marriage of Two Process Worlds. In: EuroSPI 2008, Software Process Improvement and Practice. Wiley InterScience, Hoboken (2009)
3. ISO/FDIS 26262. Road Vehicles – Functional Safety
4. ISO/IEC DTR 15504-10. Safety Extensions

# Challenges for Requirements Development: An Industry Perspective

Sandra Kelly, Frank Keenan, and Fergal McCaffery

Software Technology and Research Centre, Dundalk Institute of Technology, Dublin Rd,
Dundalk, Co Louth, Ireland
sandra.kelly@dkit.ie

**Abstract.** It is well recognised that software requirements must accurately reflect the needs of a target domain, however, challenges still exist in effectively acheiving this. This paper reports on the results of an industry-based study investigating factors that affect the communication of requirements between the development team and other stakeholders during requirements development. Challenges found in practice are related to common obstacles reported in the literature. The paper concludes with a discussion of the findings including implications for Software Process Improvment (SPI) in requirements development.

## 1 Introduction

The importance of accurately identifying software requirements is well documented. However, despite the wide range of techniques available, numerous reports indicate that successful requirements development faces many challenges. This work focuses on one important area, the relationship between developers and multiple diverse stakeholders whose collective business need must be translated into a software solution.

In order to establish the state of the art for requirements work as experienced in the development of bespoke software, this presentation will report on the findings from an industry-based study. One objective of the study is to investigate the factors that affect the communication of requirements between the development team and other stakeholders. Accordingly, attention turns to examining how the role of the customer is implemented in practice. Section two provides a list of challenges for requirements development as identified in the literature, section three details the industry study and section four provides a discussion on the findings to date including implications for software process improvement relevant to the area of requirements development.

## 2 Challenges for Requirements Development

Figure 1 shows a list of challenges for requirements development reported in the literature. Columns on the right indicate the authors of those challenges including where these overlap. For instance, four authors report obstacles identifying appropriate stakeholders and all authors list conflicting priorities as a challenge.

| Challenges | | Authors | Pressman, R.S. | Grünbacher, P. | Nuseibeh & Easterbrook | Sommerville, I. | Ambler, S. | McBryan, T. |
|---|---|---|---|---|---|---|---|---|
| 1 | Identifying appropriate stakeholders | | | | ✓ | ✓ | ✓ | ✓ |
| 2 | Limited access to stakeholders | | | ✓ | | ✓ | | |
| 3 | Stakeholders unsure of what is needed | | ✓ | | | ✓ | ✓ | |
| 5 | Conflicting priorities | | ✓ | ✓ | ✓ | ✓ | ✓ | ✓ |
| 6 | Stakeholders unable to see beyond the current situation | | | | | ✓ | | |
| 7 | Requirements change | | ✓ | ✓ | | ✓ | | ✓ |
| 8 | Stakeholders have difficulty articulating requirements | | ✓ | | ✓ | ✓ | | |
| 9 | Developers dont have implicit knowledge of the domain | | | | | ✓ | ✓ | |
| 10 | Multidisciplinarity of developers and other stakeholders | | | ✓ | ✓ | | | |

**Fig. 1.** Common Challenges and Respective Authors Reported in Literature

## 3 Industry Study Background

To date four participants have contributed to this study. These include three experts. The first was a business analyst with eleven years experience in requirements analysis. Second, a CEO with eleven years experience in requirements engineering. Third, a CEO with twenty one years experience in requirements engineering and the fourth participant had five years experience testing requirements. The experience of this person is included since analysis of test criteria involved direct contact between the participant and stakeholder groups in eliciting and clarifying requirements. Data were collected via interviews based on the experience of these practitioners. Overall, participants had worked in seven organisations developing requirements specifically for bespoke software. Two were large enterprise, five were small-to-medium enterprise and specific markets in medical/healthcare including pharmaceuticals, financial and transportation sectors.

### 3.1 Feedback

One frequently reported problem is that of not having access to sufficiently knowledgeable and authoritative stakeholders. This leads to a mismatch between perceived and actual business needs. Participants articulated a need to improvise to establish how the activities being examined are *actually* performed to ensure requirements elicited more accurately reflect the underlying business. Across all organizations identification of appropriate stakeholders involved referrals and participants described getting buy-in from relevant stakeholders as unsystematic and largely intuitive

All participants had experienced the challenge of developers not having sufficient knowledge about the problem domain. Two experts had learned to invest time understanding the target business, participating in everyday work activities and interacting with employees in the target organization for up to a week before making any decision about requirements. One stated it was important for the success of the system-to- be to discover from a software engineering perspective what was really needed, in doing so, this participatory activity had enabled both these experts to improve their knowledge about the problem domain. All participants reported that in communicating requirements, stakeholders from the target domain are unable to see beyond their own perspective of a requirement. Similar to stakeholders being unable to see beyond the current situation in section 2, this problem related to stakeholders having a narrow view of the current situation. Here, it was important that individual requirements were proactively aligned with wider and future business needs where possible. Two experts reported that in dealing with traditional formal organizations, limited access to the relevant stakeholders had been experienced.

### 3.2 Summary of Feedback

Experience in overcoming these problems has suggested solutions. For example, the two experts mentioned in section 3.1 described qualities needed to perform the customer role successfully. Here it is considered key that customer representatives are trusted within their organization, have a good understanding of their own domain, a technology oriented background and good communication skill. Overall successful communication between stakeholders is predominantly experienced when dealing with smaller groups, also face-to-face communication is preferred over electronic means for dispersed stakeholders. Group sessions such as meetings are favored techniques, with mock-ups and storyboarding useful to facilitate problem exploration in an inclusive and collaborative manner

## 4 Discussion

An interesting observation relates to the solutions of the challenges found in industry. Despite the fact that all participants had found solutions in practice to the identification and inclusion of relevant stakeholders, none of the participants had perceived their solution to be part of any formal process. However, clearly informal process pursued by each participant had contributed to overcoming these challenges. Also, solutions found involve a set of circumstances and constraints pertaining to particular situations encountered. For instance, in this work, factors that affect communication of requirements include skills and qualities possessed by the customer representative, improvisation and adaptation depending on each new situation and successfully identifying and involving relevant stakeholders is described as unsystematic. Here, manipulating a set of unpredictable factors helped to achieve success.

This work has implications for SPI since with maturity, assessment models expect to find stability through predictable requirements attributes as inputs in order to produce measurable work products as outputs. Inputs for determining capability are exclusively derived from process as documented, however, this study has found evidence indicative of bespoke systems development in industry as emergent, embracing choice and expecting every system development method to be dependent on a unique set of circumstances and constraints. A point of concern is that informal processes that prove helpful in practice are generally not documented.

It is anticipated that this work will provide further context to informal practice in requirements development and help to inform developers of assessment models in accommodating less formal yet successful practices from industry.

## Acknowledgement

This research is supported by the Science Foundation Ireland (SFI) Stokes Lectureship Programme, grant number 07/SK/I1299, the SFI Principal Investigator Programme, grant number 08/IN.1/I2030 (the funding of this project was awarded by Science Foundation Ireland under a co-funding initiative by the Irish Government and European Regional Development Fund), and supported in part by Lero - the Irish Software Engineering Research Centre (http://www.lero.ie) grant 03/CE2/I303_1.

## References

1. Ambysoft Inc., Agile Requirements Modeling (2009), http://www.agilemodeling.com/essays/agileRequirements.htm#Challenges
2. Nuseibeh, B., Easterbrook, S.: Requirements Engineering: A Roadmap. In: ICSE-2000. ACM Press, Limerick (2000)
3. Sommerville, I.: Software Engineering, 8th edn. Addison-Wesley, Reading (2007)
4. Pressman, R.S.: Software Engineering A Practitioners Approach, 5th edn., p. 252. McGraw-Hill, New York (2000)
5. Grünbacher, P.: Requirements Engineering for Web Applications. In: Kappel, G., Proll, B., Reich, S., Retschitzegger, W. (eds.) Web Engineering. John Wiley & Sons, Chichester (2006)
6. McBryan, T., McGee-Lennon, M.R., Gray, P.: An integrated approach to supporting interaction evolution in home care systems. In: 1st international Conference on Pervasive Technologies Related To Assistive Environments, vol. 282, pp. 1–8. ACM, New York (2008)

# Software Engineering Strategies: Aligning Software Process Improvement with Strategic Goals

Reinhold Plösch[1], Gustav Pomberger[1], and Fritz Stallinger[2]

[1] Kepler University Linz, Institut für Wirtschaftsinformatik - Software Engineering,
Altenberger Straße 69, 4040 Linz, Austria
{reinhold.ploesch,gustav.pomberger}@jku.at
[2] Software Competence Center Hagenberg, Softwarepark 21, 4232 Hagenberg, Austria
fritz.stallinger@scch.at

**Abstract.** Aligning software process improvement with the business and strategic goals of an enterprise is a key success factor for process improvement. Software process improvement methods typically only provide little or generic guidance for goal centered process improvements. We provide a framework for developing software engineering strategies that are aligned with corporate strategies and goals. Strategic objects as an important part of our framework can be directly aligned with SPICE or CMMI processes. This allows that any process improvement action can be systematically aligned with strategic goals.

**Keywords:** Software process improvement, software engineering strategy, functional strategy, strategic goal, CMMI.

## 1 Introduction and Overview

Aligning software process improvement with the business and strategic goals of an enterprise is a key success factor for process improvement. Intensive research has been performed on defining best practice models for software lifecycle activities (e.g. [1], [2]) as well as methods for guiding software process improvements, ranging from guidance for single improvement actions (e.g. [3], [4]) to the management of overall improvement programs (e.g.[5], [6]). Although these methods generally consider the existence of strategic goals and stress the importance of aligning process improvements to business goals, they generally provide little and typically only generic guidance on how to define the details of, prioritize and select process improvements.

In the remainder of this paper we present a method for development of functional software engineering strategies as a mediator between business goals and software process improvement. The main results presented are (a) an understanding of the role of engineering strategies in the overall strategy development context of an organization, (b) a meta-model for describing engineering strategies, (c) the identification of the strategy objects relevant for software engineering and their mapping to CMMI process areas, and (d) a method to guide the development of engineering strategies.

## 2 Developing Software Engineering Strategies

In order to understand strategy development at the software engineering level we first relate software engineering strategies to the overall strategy development efforts in an organization. Fig. 1 illustrates the overall strategy development process of an organization. We will not discuss all the steps but refer to [7] and [8] for a detailed discussion. According to [8], typically a distinction is made between the corporate strategy, various division strategies, and various functional strategies.

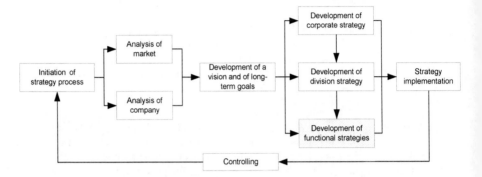

**Fig. 1.** Overall strategy development process

*Corporate Strategy:* The central issue on this level is to determine which market segments should be addressed with which resources. This has to be understood against the background of the core tasks of a company – resource allocation, diversification decisions and the coordination of the more or less independent divisions (management of synergies).

*Division Strategy:* The division strategy refines the corporate strategy. The major questions to be addressed by the division strategy are how to develop a long-term unique selling proposition compared to the market competitors and how to develop a unique product or service. The competitive advantages of a division are related to its capabilities and resources and to the customer needs and structures of the market.

*Functional Strategy:* Functional strategies define the principles for the functional areas of a division in accordance with the division strategy and therefore refine the division strategy in the distinct functional areas. Examples of such functional areas are marketing, finance, human resources, engineering, software development, etc.

In principal these strategies can be developed independently from each other; nevertheless they must all adhere to the division strategy and therefore also to the corporate strategy. While on the corporate and on the division level the emphasis is on the *effectiveness* (doing the right things) of the corporation or division, the functional strategies have their focus on the *efficiency* (doing the things right) of the respective functional areas. This distinction between the different kinds of strategies ensures that business goals are translated from the corporate strategy to the functional strategies.

In a next step we need to understand the structure of functional software engineering strategies. Fig.2 depicts the conceptual framework for the description of functional strategies. The strategic goals formulated in the software engineering strategy are refinements of strategic goals on the corporate and respectively divisional level, mapped on the functional area. A strategy object is a topic (e.g. architecturemanagement) that refines one or more strategic goals. As the strategy objects—and therefore also the strategic statements—are targeted towards the functional strategic goals it is also assured that the divisional or corporate goals are not violated.

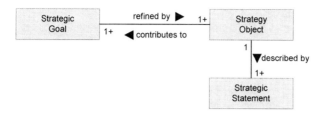

**Fig. 2.** Strategy description - conceptual framework

Each strategic goal has a description and explanation of the strategic goal, lists expected benefits, describes how to reach the strategic goal and contains a description of how to measure its realization. Furthermore, each strategic goal is prioritized. Table 1 provides an example description of a strategic goal from a real-world project. The verbalization of strategic goals is not an easy task and should be based on knowledge from a detailed analysis of the organization. A strategic goal of a software engineering strategy must not violate corporate or division goals or visions.

**Table 1.** Example for description of a strategic goal

| ID: | G-SALE | Priority: | A-Goal |
|---|---|---|---|
| Strategic Goal: | Selected software products have to be sellable separately, i.e. without selling the underlying hardware product. | | |
| Explanation of strategic goal: | The selected software products must meet conditions, so that they can be sold independently of other products (hardware and software) on the automation market. | | |
| Description how to reach the strategic goal: | This is achieved by appropriate abstraction of the runtime environment, isolation and independence from other products, extensive tests, appropriate actions for the protection of intellectual property, documentation and consulting and support offers. | | |
| Description how to measure the realization of the strategic goal: | Guideline for achieving this goal is that by the end of the first quarter of 2011 product X is sellable alone and independently of other products. | | |

Examples of strategy objects that are typically refined during the strategy development process include architecture management, quality management, requirements management, standards management, etc. The description of strategy objects comprises their definition, identification of typical topics dealt with, and examples of strategic statements. Table 2 gives an example of a description of a strategy object from a real-world project.

**Table 2.** Example for description of a strategy object

| ID: | O-WORK |
|---|---|
| Definition: | Work Organization is the systematic arrangement of effective and efficient software development and project execution. |
| Strategic statement 1: | In the areas of Firmware (incl. Technology), Human-Machine-Interface and Tools the following developer teams have to be formed: OEM development, product development, and maintenance |
| Strategic statement 2: | Each software developer is member of one of these teams. For capacity reasons a developer may temporarily join another team, but the number of these developers should be kept low. |

The general approach of the method for the systematic development of software engineering strategies is to conduct a strategy development process as shown in Fig. 3. The typical process is structured into the development and prioritization of strategic goals, strategy objects and strategic statements.

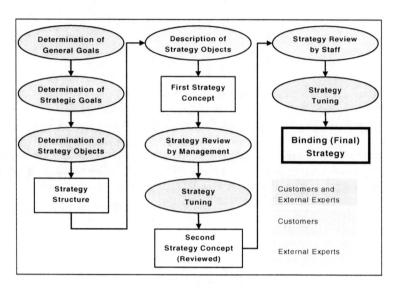

**Fig. 3.** Process of engineering strategy development

**Table 3.** Mapping of strategy objects to CMMI Process Areas

| CMMI Process Area | Strategy Object(s) | |
|---|---|---|
| Causal Analysis and Resolution | Quality Management | P |
| Configuration Management | Configuration Management | F |
| Decision Analysis and Resolution | Architecture Management, Component Management, Innovations Management | P |
| Integrated Project Management | Project Management, Work Organization | F |
| Measurement and Analysis | Quality Management, Process Management | P |
| Organizational Innovation and Deployment | Innovation Management | F |
| Organizational Process Definition | Process Management, Work Organization | F |
| Organizational Process Focus | Process Management, Work Organization | F |
| Organizational Process Performance | Process Management, Quality Management | F |
| Organizational Training | Project Management, Work Organization | F |
| Product Integration | Component Management, Quality Management, Test Management | L |
| Project Monitoring and Control | Project Management | F |
| Project Planning | Project Management | F |
| Process and Product Quality Assurance | Quality Management | L |
| Quantitative Project Management | Quality Management, Process Management | P |
| Requirements Development | Component Management, Product, Product Management, Domain Engineering | L |

**Table 3.** (*continued*)

| | | |
|---|---|---|
| Requirements Management | Requirements Management, Change Management | F |
| Risk Management | Risk Management | F |
| Supplier Agreement Management | - | N |
| Technical Solution | Architecture Management, Component Management, Domain Engineering | L |
| Validation | Quality Management | L |
| Verification | Quality Management, Test Management | F |

From our experience in applying the strategy development process, we identified a large number of strategy objects that are typically of interest for organizations. We therefore mapped the CMMI process areas to the strategy objects in order to facilitate that process improvements can be systematically aligned with the strategic specifications. The result (cf. Table 3) shows that most process areas are semantically connected with the strategy objects of our approach. This gives us the possibility to systematically crosscheck, whether process improvements are aligned with strategic decisions and goals. The mapping defines in more detail which process areas (i.e., improvements of a process area) have to be aligned with which parts of the functional software engineering strategy. The mapping is quantified by an N-P-L-F-scale adapted from [1].

## 3 Experience and Conclusion

From our experience in applying the strategy development process in various software development organizations (with more than 100 software developers, each) using our conceptual framework of strategic goals, strategy objects, etc., we can draw the following conclusions: (a) the structure helps focusing on the strategic goals; (b) as the strategy objects are linked to the strategic goals, the strategic statements are automatically targeted towards the goals of an organization; (c) the mapping of the CMMI process areas to strategy objects allows aligning identified improvements to the strategy objects and therefore (by means of the link of the strategy objects to the strategic goals) to the business and strategic goals of an organization.

## References

1. ISO/IEC 15504:2003. Information Technology - Process Assessment. International Standards Organization (2003)
2. CMMI for Development, Version 1.2. Technical Report CMU/SEI-2006-TR-008. Software Engineering Institute, Carnegie Mellon University, Pittsburgh, PA (2006)
3. Shewhart, W.A.: Economic control of quality of manufactured product. D.Van Nostrand Company, New York (1931)
4. Dion, R.: Process improvement and the corporate balance sheet. IEEE Software 10(4), 28–35 (1993)

5. McFeeley, B.: IDEAL: A user's guide for software process improvement. Handbook CMU/SEI-96-HB-001, Software Engineering Institute, Carnegie Mellon University, Pittsburgh (1996)
6. ISO/IEC 15504-7:1998. Information Technology - Software Process Assessment - Part 7: Guide for use in process improvement. International Standards Organization (1998)
7. Simon, H., von der Gathen, A.: Das große Handbuch der Strategieinstrumente – Alle Werkzeuge für eine erfolgreiche Unternehmensführung. Campus, Frankfurt/Main (2002)
8. Venzin, M., Rasner, C., Mahnke, V.: Der Strategieprozess – Praxishandbuch zur Umsetzung im Unternehmen. Campus, Frankfurt/Main (2003)

# Deploying Lifecycle Profiles for Very Small Entities: An Early Stage Industry View

Rory V. O'Connor [1,2] and Claude Y. Laporte [3]

[1] Lero, the Irish Software Engineering Research Centre, Ireland
[2] Dublin City University, Dublin, Ireland
[3] École de technologie supérieure, Montréal, Canada
Rory.OConnor@computing.dcu.ie, Claude.Y.Laporte@etsmtl.ca

**Abstract.** The recently published ISO/IEC 29110 standard Lifecycle profiles for Very Small Entities has at its core a Management and Engineering Guide [1] which is targeted at very small entities (enterprises, organizations, departments or projects) having up to 25 people [2], to assist them unlock the potential benefits of using standards which are specifically designed to address their needs. This paper will also outline this standard and the implementation of a series of pilot project initiative harnessing a set of detailed guidelines known as "Deployment Packages" to assist very small entities in understanding and exploring the potential usage of an international software process standard. This paper will address issues of small entities needs, industry reaction to early pilot projects and highlight the needs for a light weight process assessment mechanism to meet the needs of very small entities and complement this new lifecycle standard.

**Keywords:** VSE, ISO/IEC 29110, Standards.

## 1 Introduction

In a time when software quality is a key to competitive advantage, the use of ISO/IEC systems and software engineering standards remains limited to a few of the most popular ones. Research shows that Very Small Entities (VSEs) can find it difficult to relate ISO/IEC standards to their business needs and to justify the application of the standards to their business practices [2, 3]. Most of these VSEs don't have the expertise or can't afford the resources - in number of employees, cost, and time - or see a net benefit in establishing software life-cycle processes. There is sometimes a disconnect between the short-term vision of the company, looking at what will keep it in business for another six months or so, and the long-term benefits of gradually improving the ways the company can manage its software development and maintenance. A primary reason cited by many small software companies for this lack of adoption of such ISO standards, is the perception that they have been developed for large software companies and not with the small organisation in mind [3]. Subsequently, VSEs have no or very limited ways to be recognized as enterprises that produce quality software systems in their domain and may therefore be cut off from some economic activities.

## 2 ISO/IEC 29110 Background

Accordingly there is a need to help such organizations understand and use the concepts, processes and practices proposed in the ISO/IEC JTC1/SC7's international software engineering standards. The ISO/IEC 29110 standard "Lifecycle profiles for Very Small Entities" [1] is aimed at addressing the issues identified above and addresses the specific needs of VSEs [2].

The approach [2] used to develop ISO/IEC 29110 started with the pre-existing international standard ISO/IEC 12207 [4] dedicated to software process lifecycles. The overall approach consisted of three steps: (1) Selecting ISO/IEC 12207 process subset applicable to VSEs of less than 25 employees; (2) Tailor the subset to fit VSE needs; and (3) Develop guidelines for VSEs.

At the core of this standard is a Management and Engineering Guides (ISO/IEC 29110-5) [1] focusing on Project Management and Software Implementation.

The core characteristic of the entities targeted by ISO/IEC 29110 is size, however there are other aspects and characteristics of VSEs that may affect profile preparation or selection, such as: Business Models (commercial, contracting, in-house development, etc.); Situational factors (such as criticality, uncertainty environment, etc.); and Risk Levels. Creating one profile for each possible combination of values of the various dimensions introduced above would result in an unmanageable set of profiles. Accordingly VSE's profiles are grouped in such a way as to be applicable to more than one category.

Profile Groups are a collection of profiles which are related either by composition of processes (i.e. activities, tasks), or by capability level, or both. The "Generic" profile group has been defined [1] as applicable to a vast majority of VSEs that do not develop critical software and have typical situational factors. This profile group does not imply any specific application domain, however, it is envisaged that in the future new domain-specific sub-profiles may be developed in the future.

To date the Basic Profile standard has been published by ISO, the purpose of which is to define a software development and project management guide for a subset of processes and outcomes appropriate for characteristics and needs of VSEs developing a single application. Work is current underway on the Entry and Intermediate profiles. VSEs targeted by the Entry Profile are VSEs working on small projects (e.g. at most six person-months effort) and for start-up VSEs while VSEs targeted by the intermediate profile are developing multiple applications.

## 3 Deployment Assistance

The issues of assistance to VSEs in understanding and adopting standards, as outlined in section 1, must be addressed. To this end, some members of the ISO/IEC JTC1/SC7 WG 24 have produced a set of "Deployment Packages" (DP). A DP is a set of artifacts developed to facilitate the implementation of a set of practices, of the selected framework, in a VSE. A DP is not a process reference model (i.e. it is not prescriptive). The elements of a typical DP are: description of processes, activities, tasks, roles and products, template, checklist, example, reference and mapping to standards and models, and a list of tools. The mapping is only given as information to show that a deployment package has explicit links to standards, such as ISO/IEC 12207, or models, such as the CMMI for Development, hence by deploying and implementing the package, a VSE can see its concrete step to achieve or demonstrate coverage. Packages are designed such that a VSE can implement its content, without having to implement the complete framework at the same time. These DPs are freely available from [5]:

In addition a series of "Implementation Guides" have been developed to help implement a specific process supported by a tool and are freely available from [5]. To date a small number of implementation guides have been developed. These include: Version Control with CVS; Version Control with SVN; and Project Management with GForge.

## 4 Pilot Implementation Projects

The working group (ISO/IEC JTC1/SC7 WG 24) behind the development of this standard is advocating the use of pilot projects as a mean to accelerate the adoption and utilization of ISO/IEC 29110 by VSEs. Pilot projects are an important mean of reducing risks and learning more about the organizational and technical issues associated with the deployment of new software engineering practices. A successful pilot project is also an effective means of building adoption of new practices by members of a VSE. Pilot projects are based on the ISO/IEC 29110-5 Management and Engineering Guide [1] and the deployment package(s). In particular these are aimed to collect, as a minimum, the following data:

- Effort and time to deploy by the VSE
- Usefulness for the VSE
- Verification of the understanding of the VSE
- Self-assessments data - A self-assessment at the beginning of the pilot and at the end of the pilot project DP

To date a series of pilot projects have been completed in several countries utilizing some of the deployment packages developed. For example in Canada a pilot study has been conducted with an IT department with a staff of 4: 1 analyst and 3 developers, who were involved in the translation and implemented 3 DPs: Software Requirements, Version Control, Project Management. In Belgium a VSE of 25 people started with a process assessment phase aiming to identify strengths and weaknesses in development related processes. This company is now working on improvement actions mainly based on the following Deployment Packages: Requirement Analysis, Version Control, and Project Management. In France, a pilot study [6] was conducted with a 14-people VSE that builds and sells counting systems about the frequenting of natural spaces and public sites. In addition a further series of pilot projects are currently underway in Canada, Ireland, Belgium and France, with further pilot projects planned in the near future.

## 5 Discussion

As ISO/IEC 29110 is an emerging standard there is much work yet to be completed. The main remaining work item is to finalise the development of the remaining three profiles: (a) Entry - six person-months effort or start-up VSEs; (b) Intermediate - Management of more than one project and (c) Advanced - business management and portfolio management practices.

With any new initiative there is much to be learnt from conducting pilot projects. One issue of major importance to VSEs which is emerging from these pilot projects and similar work by the ISO working group is the need for a light-weight flexible approach to process assessment. Whilst work is currently underway on an assessment mechanism for ISO/IEC 29110 [7], a clear niche market need is emerging which may force the process assessment community to change their views on how process assessments are carried out for VSEs. In particular there is a strong need to ensure that VSEs are not required to invest the anything similar in terms of time, money and other resources on process assessments, as may be expected from there larger SME (small and medium enterprises), or even MNC (multinational corporation) counterparts. Indeed some form of self-assessment, possibly supported by Internet based tools, along with periodic spot-checks may be suitable alternative to meet the unique needs of VSEs. It is clear that the process assessment community will have to rethink process assessment, new methods and ideas for assessing processes in VSEs.

**Acknowledgments.** This work is supported, in part, by Science Foundation Ireland grant 03/CE2/I303_1 to Lero, the Irish Software Engineering Research Centre (www.lero.ie).

## References

1. International Organization for Standardization (ISO). ISO/IEC TR 29110-5-1-2 Software Engineering - Lifecycle Profiles for Very Small Entities (VSEs) – Management and Engineering guide: Generic profile group, Basic Profile, Geneva (2011)
2. Laporte, C.Y., Alexandre, S., O'Connor, R.: A Software Engineering Lifecycle Standard for Very Small Enterprises. In: O'Connor, R., et al. (eds.) Proceedings of EuroSPI. CCIS, vol. 16, pp. 129–141. Springer, Heidelberg (2008)
3. Coleman, G., O'Connor, R.: Investigating Software Process in Practice: A Grounded Theory Perspective. Journal of Systems and Software 81(5), 772–784 (2008)
4. ISO/IEC 12207. Information technology – Software life cycle processes. International Organization for Standardization/International Electrotechnical Commission, Geneva, Switzerland (2008)
5. ISO/IEC JCT1/SC7 Working Group 24 Deployment Packages repository, http://profs.logti.etsmtl.ca/claporte/English/VSE/index.html
6. Ribaud, V., Saliou, P., O'Connor, R., Laporte, C.: Software Engineering Support Activities for Very Small Entities. In: Riel, et al. (eds.) Systems, Software and Services Process Improvement. CCIS, vol. 99, pp. 165–176. Springer-Verlag, Heidelberg (2010)
7. International Organization for Standardization (ISO). ISO/IEC TR 29110-3, Software Engineering - Lifecycle Profiles for Very Small Entities (VSE) - Part 3: Assessment Guide. Geneva (2011)

# Past, Present and Future of Process Improvement in Ireland – An Industry View

Fran O'Hara

Inspire Quality Services, Dublin, Ireland
fran.ohara@inspireqs.ie

**Abstract.** This experience report relates to the authors experiences of over 15 years of being involved in process improvement in Ireland with a wide variety of organisations. Experiences relating to process assessments and to approaches to improve the process will be shared. This includes software process improvement as well as test process improvement and observations on the relationship between them. The author's view of the present and indeed future need for pragmatic and rapid process improvement and how this relates to the increasing use of agile methods is discussed.

## 1 Introduction

Over the last 15 years or so the author has been involved in a variety of process assessments and also supporting improvements both in general Software Process Improvement (SPI) and also test process improvement (TPI). In Ireland, the organizations embarking on process improvement range from the indigenous SMEs to the large multi-national organizations and within sectors ranging from banking and financial services to ICT and product companies. Motivation for process improvement programmes typically range from corporate directives, to marketing (internally and externally leveraging process maturity 'ratings') to business focused performance improvement. Over this period, there were also a number of EU and Irish Government funded initiatives to support focused improvements, increase awareness of SPI, etc.

From an assessment perspective, during this 15 year period the author has been involved in areas such as

1. performing formal and informal software/systems process assessments using models such as CMMI® and SPICE®
2. performing formal and informal test process assessments using models such as TPI® and TMMi®
3. performing 'consultancy' assessments and benchmarking exercises without the explicit use of a model

The next section will present the highlights relating to these process assessment experiences and what are learning points specific to Ireland versus those that are more widely applicable.

However, a more significant portion of the author's experience has also been in supporting organizations in the actual implementation of improvements (some based on formal assessment/improvement frameworks others not) and again the key learning points will be covered in the next section.

Note: all views expressed are based on the author's subjective experiences and is an industry view rather than a formally researched analysis.

## 2 Past

One of the early involvements was the EU funded SPIRE [1] project where I was involved in one of a number of small focused facilitated self-assessment and improvement projects based on SPICE acting as a mentor. The focus area was the requirements area and the organization gained measurable benefits from the incremental improvement and published a case study. The learning point though was the difficulty in sustaining the improvement by tackling other process areas within the organization after the initial pilot.

In general, in the author's experience, the CMM/CMMI® has been the more widely used process improvement model in Ireland. Larger organizations, particularly the multi-nationals, tended to adopt formal programmes of improvement with formal assessments. Some used staged models in pursuit of maturity levels while other used continuous models when wanting to focus more specifically on business performance goals. Smaller organizations tended to use the models as 'toolboxes' of ideas to help them improve rather than following the models formally in a structured assessment and improvement programme. In general, these improvement programmes had a largely positive impact but with some mixed results. Initiatives such as Lero helped particularly with practical improvements in SMEs. The cost of formal CMMI® based improvement and particularly formal assessments was one of its inhibitors as was evidenced by example with the limited lifetime of the Northern Ireland initiative on CMMI® adoption.

On the test model side, TPI® has been more widely used that TMM/TMMi® as it is only recently that a standard detailed reference model became available for the TMMi® (with a non-profit organization, the TMMi Foundation, behind it to promote it as an industry standard).

In general, the Irish context is one that necessitated an intensely practical and often tailored or non-standard approach. This is due in no small part to the large portion of Small to Medium Enterprises in the Ireland but it is also partly a cultural response. Based in part on this requirement for a more rapid focused improvement approach, the author co-developed an incremental process improvement method called 'Rapid Performance Improvement' (RPI®) which proved effective in supporting process improvements particularly where there was a need for prioritization and fast results in support of business drivers. A number of tutorials and case studies on RPI have been presented at process improvement and testing conferences over the years (e.g. [2,3]). The key learning points from the development and use of this approach were:

- Stakeholder buy-in (both management and staff) benefitted greatly from rapid incremental improvements delivering tangible results rather than long cycles of improvement
- Management must see fast measurable results that impact on their key business drivers. The link between process improvements and prioritized business goals and development goals must be made as explicit and measurable as possible.

- Staff must see the relevance of process improvement to them and their work on projects. It must solve the day to day problems that they face on projects. Perception is hugely important for success when managing change.

Another interesting experience was the interaction between software/systems process improvement and test process improvement. There are two quite distinct communities underpinned by the separate roles in organizations – one involving test professionals that focuses on test process improvement and the other more development/project management professional community involved in SPI in general (and of which test is of course a part as well). Yet there are both commonalities and key integration learning points that can be deduced from these experiences that should be shared to a greater extent. Not least of these is the need for prioritization between the improvement areas across test and software/systems process improvement to maximize support for business drivers. There is a diminishing return to be had from improving the testing process when constraints from project management, software development processes and organizational structures mean that scarce resources would be better spent improving those areas with more significant benefit. The benefit of broader software/systems models like SPICE® and CMMI® is of course that they provide this more holistic picture. The advantage of the test models like TPI® and TMMi® is the greater detail they provide in the testing area. A logical conclusion is to combine software process improvement with test process improvement to get the benefits of both. An integrated approach to the prioritization of areas for improvement in support of goals and problems such as that included in the RPI® method helps achieve this prioritization/integration.

## 3 Present

In Ireland there is also now an increasing use of agile practices which is raising many issues and questions about the application of process improvement concepts and models with an increasingly agile context. Some organizations are 'embracing the new' and adopting agile methods while walking away from the practices that have been effective in the past in their context. This can be short-sighted particularly because many agile methods are not meant to be fully encompassing of the practices an agile project will need.

There is a significant cultural change associated with agile that is required to make it deliver on the predictability and productivity improvements espoused by advocates. Plan-driven approaches can hinder a shift to self-directed teams if the organization is not careful. Partial agile implementations which do not embrace key elements of an agile culture such as empowerment, or partial implementation of practices from an agile method can also reduce chances of success with agile. Extreme programming for example is a very disciplined agile approach but many organizations particularly in Ireland tailor methods to their context. Sometimes, however, this 'tailoring' means simply omitting practices without full understanding of implications. An example of this would be the common omission of pair programming in Extreme Programming without compensating with alternative practices - quality levels are then typically negatively impacted.

A benefit of considering and interpreting traditional process improvement models in an agile context is that they can help reflect on the 'big picture' to help achieve a balanced implementation.

Scrum has rapidly become the most popular agile method in Ireland and indeed elsewhere. However it is a relatively straightforward work management approach that includes nothing about specific development and testing practices. Projects must define those practices that will work best within this framework in order for the Scrum to work effectively in their project/organisation. A mixture of existing practices from the organization combined with practices such as Test Driven Development from Extreme Programming are what the author often sees working well in organizations.

Most proponents of process improvement models will declare that the models can be used with agile. The Software Engineering Institute for example has published many articles to this effect re the CMMI and agile. The difficulty is in the fact that one must significantly interpret the models and indeed find 'agile aware' assessors to combine traditional process improvement with agile methods. There is an increasing number of organizations who are doing this successfully although case studies in Ireland are still rare. In Ireland, a significant portion of the organizations adopting agile methods that the author is aware of, are adopting hybrid combinations that for example

- Mix or combine agile methods such as Scrum with practices from Extreme Programming
- Mix plan-driven and processes aligned with terminology in the traditional models with agile methods. An example would be when organisations need to do fixed price proposals they rely on their plan-driven practices with structured metrics based estimation processes but then work in iterations using methods like Scrum and practices such as story point estimation with 'planning poker' (similar to the Delphi technique) and the concept of 'velocity' to utilise feedback on the measured productivity of the team from iteration to iteration.
- Mix agile with traditional practices because of their specific context such as regulated environments and the need to support auditing.

## 4 Future

In the short to medium term it is reasonable to expect that traditional process improvement will continue to evolve in support of industry requirements. Key areas for this include

- continuing adaptation of models and approaches in support of specific industry domains such as regulated environments (e.g Medi SPICE), automotive, etc.
- the newer test models/adaptations will most likely achieve increasing adoption. This includes the TMMi®, TestSPICE, the updated TPI® Next and so on.
- greater uptake of approaches to support rapid performance based improvement involving short cycles of change (driven in part by the economic driven investment justification)

There is no doubt that agile methods will continue to become more widespread. There are already case studies being reported of heavily regulated environments using agile practices such as exploratory testing and being commended by regulatory authorities for doing so.

Some agile methods imply that the empowered teams have complete freedom to choose their own processes and tools to get the job done for their project. When taken to the extreme there is little or no reuse of good practice between teams with the wheel constantly being reinvented and a proliferation of tools across projects.

On the other hand a rigid organisational process is largely at odds with an agile culture that at its core typically includes self-directed teams with freedom to choose and therefore with real responsibility and ownership to get the job done. From the authors observation to date, the reality for many organizations adopting and using agile in the future will be one where there is some blend in between these two extreme scenarios. A supportive yet flexible organizational perspective on process and supporting toolsets that guides and facilitates rather than dictates or mandates.

With the increasing use of agile methods and based in part on the observations above, the author believes the following are key considerations moving forward:

- greater explicit support within traditional process improvement models for use with agile environments
- greater support in terms of change management approaches and implementation guidance for hybrid implementations of agile and non-agile methods

## References

1. Sanders, M.: The Spire Handbook: Better, Faster, Cheaper Software Development in Small Organization. Centre for Software Engineering, Dublin Ireland (1998)
2. Hart, J., O'Hara, F.: How to rapidly define the organization's standard set of processes Tutorial presented at SEPG (2005)
3. Frazer, M., O'Hara, F.: Experiences with a risk-based approach to CMMI implementation. presented at the European SEPG (2005)

# Neural Network Based Effort Prediction Model for Maintenance Projects

V. Bharathi and Udaya Shastry

Wipro Technologies, 53/1, Ganapa Towers, Madiwala, Bangalore, 560068 India
{bharathi.kumar,udaya.shastry}@wipro.com

**Abstract.** One of the most critical requirements of High Maturity practices is the development of valid and usable prediction models (Process Performance Model, PPM) for quantitatively managing the outcome of a process. Multiple Regression Analysis is a tool generally used for model building. Over the last few years, Artificial Neural Networks have received a great deal of attention as prediction and classification tools. They have been applied successfully in diverse fields as data analysis tools. Here, we explore the applicability of neural network models for bug fix effort prediction in corrective maintenance project and present our findings

**Keywords:** Process Performance Model, Neural Networks.

## 1 Introduction

Increased emphasis on the implementation of prediction models is driven by theindustry standards E.g., CMMI [1]and Automotive SPICE® [2]. As a result, organizations are moving to the next level of quantitative management where empirical methods are used to establish process predictability, thus enabling better project planning and management.

Regression under the Least-Squares model, a most common tool employed for prediction assumes a reasonably normal underlying data distribution. However datasets derived from software engineering do not always adhere to this assumption– data is often skewed [3]. In such 'non-normal' cases the least–squares regression model loses much of its efficiency [4].

Analysis of such complex datasets can be achieved through Artificial Neural Networks (ANN). ANNs are rapidly gaining popularity as data analysis tools. With their remarkable ability to derive meaning from complex dataset, they are used to extract patterns and detect trends that are too intricate to be noticed by either humans or other computer techniques. They have been successfully applied in many fields from science [5] to engineering [6] and from management [7]to control [8]. In this paper, we are presenting our experience on applying neural networks for process prediction.

## 2 Background

We applied neural networks for the design of a PPM for effort prediction of a bug fix (CR) in a corrective maintenance project. With software maintenance accounting for an excess of 50% of the total programming effort [9], accurate effort estimation has major implications. If the estimate is too low, the project team will be under considerable pressure to deliver the product quickly, and hence the deliverable may contain residual errors. On the other hand if the estimates are on a higher side, more number of resources needs to be committed for the project, adding to the cost.

## 3 Methodology

All the data used in this study belong to Telematics area of Automotive domain. Total effort required for a bug fix was chosen as a response. Attributes like effort for reproduction, knowledge level of the developer, code complexity, changes to design, dependency on other modules, testing effort and impact on the base acted as inputs. For modeling the above data, we selected one hidden layer Multilayer Perceptron (MLP) feed forward architecture with back propagation training algorithm.

The neural network structure is realized using NeuroSolutions software version 6.04. A learning rate of 0.5, a momentum rate of 0.7 and random initial weights are chosen. The training was stopped when the cross validation error began to increase as this is considered to be the point of best generalization.

Once the network was trained and the weights were frozen, the saved neural network was applied to the test data. The results of the training, cross validation and testing are discussed in detail in the next section.

## 4 Results and Discussions

The figures 1, 2 and 3 show the learning curves for both the training and cross validation data for the incremental datasets A(100 samples), B(150 samples) and C(250+ samples). Training was stopped when mean square of the estimation error(MSE) of cross validation began to increase indicating over-fitting.

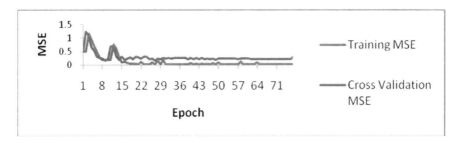

**Fig. 1.** Learning curve for dataset A

However, the true test of a network is how well it can perform when presented with data it has not seen before. Table 1 summarizes the performance metrics for the test datasets. The "Correlation coefficient" metric indicates the fit of the model to the data. The "% correct" metric indicates the correctness of the prediction in percentage.

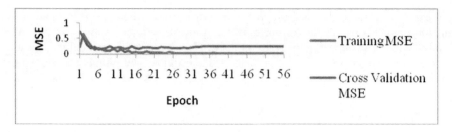

**Fig. 2.** Learning curve for dataset B

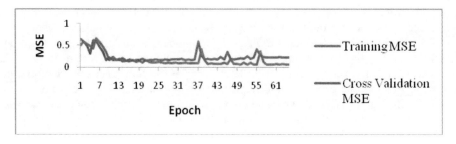

**Fig. 3.** Learning curve for dataset C

The correlation coefficient r is 0.37 for dataset A. It improved to 0.46 with dataset B and to 0.6 with dataset C. Also, the prediction accuracy obtained for the dataset A is 50%, lowest as compared to that of dataset B and C. It improved to 53% with dataset B and to 70% with dataset C.

**Table 1.** Performance metrics: Test data

|  | Dataset A | Dataset B | Dataset C |
|---|---|---|---|
| MSE | 0.224451 | 0.110763 | 0.045096 |
| Correlation (r) | 0.374946 | 0.463404 | 0.601357 |
| % Correct | 50.00% | 53.33% | 70.00% |

The model with dataset C has yielded a prediction accuracy of 70%. From practitioners' point of view, we consider that this value is good enough.

We found that neural network performed worse when few data are available. A large dataset is always desirable in neural network applications. The pattern when repeated will be valuable for the learning process.

## 5 Conclusions

In this study, we have presented an ANN based process performance model developed for bug fix effort prediction in a corrective maintenance project, using historical data. Implementation and effective application of process performance models is a key High Maturity requirement from standards like Automotive SPICE® and CMMI.

The results show that neural network can be used for effort prediction. Application of ANN technique will particularly be useful when there are implicit interactions and non-normal data.

Future goal is to develop anordinal logistic regression model for the same incremental datasets and compare the performance with ANN based model.

## References

1. CMMI Development Team. Capability Maturity Model® Integrated Version 1.2, Software Engineering Institute (2001)
2. Automotive SIG. Automotive SPICE© Process Assessment Model (v2.5) and ProcessReference Model (V4.5) (2010)
3. Kitchenham, B., Pickard, L.: Towards a constructive quality model part ii, Statistical techniques for modeling software quality in the esprit request project. Software Engineering Journal 2(4), 114–126 (1987)
4. Hampel, F.R., Ronchetti, E.M., Rousseeuw, P.J., Stahel, W.A.: Robust statistics. John Wiley & Sons, New York (1986)
5. Namdar-Khojasteh, D., Shorafa, M., Omid, M., Fazeli-Shaghani, M.: Application of Artificial Neural Networks in Modeling Soil Solution Electrical Conductivity. Soil Science 175(9), 432–437 (2010)
6. Intelligent Engineering Systems Through Artificial Neural Network. In: Proceedings of the Artificial Neural Networks in Engineering Conference, November 5-8, vol. 10. St Louis, Missouri (2000)
7. Krycha, K.A., Wagner, U.: Applications of artificial neural networks in management science: a survey. Journal of Retailing and Consumer Services 6(4), 185–203 (1999)
8. Fukuda, T., Shibata, T.: Theory and applications of neural networks for industrial control systems. IEEE Transactions on Industrial Electronics 39(6), 472–489 (1992)
9. Sarkar, S., Sindhgatta, R., Pooloth, K.: A Collaborative Platform for Application Knowledge Management in Software Maintenance Projects, Compute 2008, January 18-20, 2008, Bangalore, Karnataka, India ©. ACM, New York (2008) ISBN 978-1-59593-950-0 /08/01

# A Framework of Organizational E-Readiness Impact on E-Procurement Implementation

Naseebullah, Shuib Bin Basri, P.D.D. Dominic, and Muhammad Jehangir Khan

Department of Computer and Information Sciences Universiti Teknologi PETRONAS
Bandar Seri Iskandar, Tronoh Perak, Malaysia
naseeb_lango@yahoo.com, shuib_basri@PETRONAS.com.my,
dhanapal_d@PETRONAS.com.my, janisbg22@yahoo.com

**Abstract.** Electronic Procurement (E-procurement) is an application of Electronic Commerce (E-Commerce) that facilitates corporate purchasing over the Internet. E-procurement is expected to play a foremost role in supply chain management and improve the service of delivery. However, not every firm has been successful in E-procurement implementation; most of organizations that fail or delay the implementation are due to non organizational assessment at micro level. To address this issue there is a need for an advanced assessment of organizational electronic readiness (e-readiness) to fuel a concrete planning for the implementation of E-procurement in Malaysian based organizations. This paper attempted to explore the literature on organizational e-readiness factors that lead to E-procurement implementation. Eventually, a proposed theoretical framework is developed for organizational e-readiness that consolidates relevant factors that have been categorized into perceived management readiness, perceived technological readiness and perceived environmental readiness. Beside the impact of organizational e-readiness to E-procurement implementation, market turbulence also has been identified as a moderating factor between organizational e-readiness and E-procurement implementation.

**Keywords:** E-readiness, E-procurement implementation, framework.

## 1 Introduction

The advent of Internet has significantly changed many firm's operations and become a universal source for general public, government and business communities. Such an Information Technology (IT) creates a competitive environment among organizations at both domestically and internationally and become a global open market for everyone. Now many organizations are focusing on shifting their operation from traditional to an E-procurement and E-supply chain [1], in order to reduce costs and improve the delivery of goods and services. Procurement of goods and services for raw materials and spare parts of most industries constitute 50% to 70% for high technology firms [2].

E-procurement is defined as "A comprehensive process in which organizations use IT systems to establish agreements for the acquisition of products or services (contacting) or purchase products or services in exchange for payment (purchasing)" [3]. According to a survey data 11% to 12% business growth and 35% cost reductions has been experienced by organizations after E-procurement implementation [4].

Whilst implementing E-procurement need an advanced assessment of organization's e-readiness. E-readiness is defined here as "a measure to which an organization or business may be ready, prepared or willing to adopt, use and benefits arise from the digital economy such as e-procurement" [5]. Assessing e-readiness is significant to judge the impact of (ICTs), help to determine current situation and plan for future changes [6]. As stated by [7], the impact of e-readiness success on e-commerce is also based on readiness assessment. Foundations of e-business and e-commerce can take place only by emergent initiatives of readiness [8]. Some studies propose e-readiness model as a foundation for firm to adopt E-commerce [9]. However, a study by [10] argued that the successful adoption of E-commerce strategy depends on its perceived e-readiness in managerial, organizational and environmental contexts. As for this, it is important to investigate organization's e-readiness for E-procurement implementation.

This paper attempted to explore a comprehensive survey on existing literature. Based on literature review a proposed theoretical framework is developed for organizational e-readiness that could lead to E-procurement implementation.

## 2 Framework and Hypothesis Development

In this section, we proposed a theoretical framework for the implementation of E-procurement in Malaysia based organizations. This framework is based on the intensive literature survey in two phases. First phase is the organizational E-readiness and second phase is the E-procurement implementation. This framework (fig. 1) has been used to study the E-procurement implementation. The details of framework are discussed below.

**Perceived Management Readiness**

a) Management Awareness

Management awareness means the understanding and knowledgeable about technologies and its potential benefits to an organization. According to [10], Management awareness refers to "an understanding of e-commerce technologies, business models, requirements, benefits and threats and projection of the future trends of e-commerce and its impact". Management awareness about competitor's technology highly influences the implementation of E-procurement. Relationship of management awareness in terms of organizational e-readiness for E-procurement implementation, we could propose a hypothesis as below.

H1. Management awareness has a positive influence on organizational e-readiness for E-procurement implementation.

b) Management Support/Commitment

Top management support and commitment has been considered crucial in shaping organizational strategies and development in implementation process [11]. The implementation of technology significantly depends on managerial attention and financial commitment; such support and investments are not possible without the approval of top management [12]. In this regard top management support is essential in overcoming barriers and resistance to change [4]. Without the support and commitment from top management it is impossible to successfully implement E-procurement [13] in most of the organizations. From this positive relationship of management support, we could propose a hypothesis as below.

H2. Management support/commitment has positive influence on organizational e-readiness for E-procurement implementation

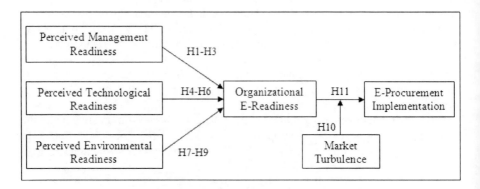

**Fig. 1.** Theoretical Framework

a) Management Financial Resources

To get survive in a competitive market, organizations must have sound enough financial resources to implement technology and technology need financial resources. Organizations with accessible computer hardware, software, technology training, user support and other implementation policies and practices incur substantial financial resources [14]. Without financial resources organization face great difficulty to implement E-procurement for their competitive advantage. Therefore, we could propose a hypothesis as below.

H3. Management financial resources has a positive influence on organizational e-readiness for E-procurement implementation

**Perceived Technological Readiness**

b) IT Infrastructure

IT infrastructure has often been identified as a successful predictor of IT adoption [15]. An efficient and effective IT infrastructure is the base of implementing E-procurement. Basically IT infrastructure means the required IT tools which may be equipment, software, hardware, systems etc significantly help in E-procurement implementation. Therefore, we could propose a hypothesis as below.

H4. IT infrastructure directly contribute to the organizational e-readiness for E-procurement implementation

c) IT Expertise

IT expertise describes firm intensity of specific knowledge and skills in the field of IT to operate day to day business transactions and activities. Firms with high levels of IT technical expertise can be expected to achieve more firm performance than firms with lower levels of technical expertise [16]. Those firms with technical expertise have more chance and easy to implement E-procurement in their organization. For this, we could propose a hypothesis as below.

H5. IT expertise directly influence the organizational e-readiness which helps in E-procurement implementation

d) Perceived Compatibility

Compatibility defines as "degree to which an innovation is perceived as being consistent with the existing values, past experiences, and needs of the potential adopter" [17]. An innovation is more likely to be adopted when it is compatible with individual's job responsibility and value systems [18]. An innovation can be adopted when it might be perceived as technically or financially superior in accomplishing a given task, but not irrelevant to its needs [17]. The compatibility of innovation is significantly influence the implementation and adoption in various studies of researchers [19]. Therefore based on such relationship, we could propose a hypothesis as below.

H6. Perceived compatibility has a positive influence on organizational e-readiness which helps in E-procurement implementation

**Perceived Environmental Readiness**

e) Customer Support/Commitment

Customer support and readiness is measured through their confidence and trust on the online transaction. Consumer's transaction on the web is mounting at a lower space than expected because consumers have a lack of trust on most sites of the internet [20]. A number of customers are still hesitant to do transaction because of the security and privacy concerns [21]. In case of implementing E-procurement customer readiness also effect on implementation. Therefore, we could propose a hypothesis as below.

H7. Customer's orientation directly contribute to the organizational e-readiness which helps in E-procurement implementation

f) Suppliers Support/Commitment

In Business-to-Business (B2B) application, suppliers are one of trading partners for information sharing, negotiation, contract for tendering and buying and selling of goods and services among organizations. In this case suppliers have a significant role for most of organizations especially the manufacturing industries. As in technology world most of organizations have direct collaboration with their trading partners (suppliers). Suppliers support and commitment has a positive significance on e-procurement implementation [22]. Therefore, we could propose a hypothesis as below.

H8. Supplier's support/commitment has directly influence the organizational e-readiness for E-procurement implementation

g) Government Policies and Regulations

Every country has its own policy and regulations to monitor and facilitate organizations [23], regarding technology adoption and making law and policies for cyber crime. Most of the countries government are much initiative and effective in promoting SMEs with financial and technology incentives for long term. The country with high level of effective policies and regulations for technology implementation is more likely to implement E-procurement in their organizations. Therefore, we could propose a hypothesis as below.

H9. Government policies and regulations has directly contribute to the organizational e-readiness for E-procurement implementation

**Market Turbulence**

Markets are often volatile and unpredictable [24], and firms face great challenges to set strategies that fit the Market characteristics [25]. In this situation, firms must be able to more aware of the turbulent market to handle business routines and make efficient decisions [26]. Such changing of technologies may affect organizational e-readiness and E-procurement implementation. Therefore, from relationship expectation above, we could propose a hypothesis as below.

H10. Market turbulence has a positive moderating influence on the relationship between organizational e-readiness and E-procurement implementation

## 3 Conclusions and Future Work

E-procurement implementation needs to assess or evaluate organizations current readiness in management, technological and environmental perspectives. Without assessing and evaluating organization's e-readiness it may cause delay or failure in implementing E-procurement. A number of factors attempted to develop a framework of organizational e-readiness on E-procurement implementation. This framework will be significant to determine the organization's capability to adopt technology.

A quantitative data will be collected from Malaysia based organizations to test the hypotheses of our framework. Once this framework is tested with the help of hypotheses, it may be used for a comparative study among developing countries.

## References

1. Lee, J., Ni, J., Koc, M.: Draft report NSF workshop on Teether free technology for e-manufacturing, e-maintenance and e-service, organized by NFS industry/University Co-operation Research Center, Wiscon Sin, USA (2001)
2. Wood, C.A.: Future and current insights from online auctions. A Research Framework of Selected Articles in Online Auctions (2004)
3. Gunasekaran, A., Ngai, W.T.E.: Adoption of e-procurement in Hong Kong: An empirical research. International Journal of Production Economics 113, 159–175 (2008)
4. Teo, T.S.H., Lin, S., Lai, K.H.: Adopters and non-adopters of e-procurement in Singapore: An empirical study. Omega 37, 972–987 (2009)
5. Lou, E.C.W., Goulding, J.S.: The pervasiveness of e-readiness in the global built environment arena. International Journal of Systems and Information Technology 12(3), 180–195 (2010)
6. Budhiraja, R., Sachdeva, S.: eReadiness Assessment. In: Conference Proceedings, International Conference on Building Effective eGovernance, Chandigarh, India (2002)
7. Molla, A.: The impact of eReadiness on eCommerce Success in Developing Countries: Firm Level Evidence. Institute for Development Policy and Management (2004) ISBN: 1 904143 482

8. Ruikar, K., Anumba, C.J., Carrillo, P.M.: VERDICT—An e-readiness assessment application for construction companies. International Journal of Automation in Construction 15, 98–110 (2006)
9. Fathiana, M., Akhavanb, P., Hooralia, M.: E-readiness assessment of non-profit ICT SMEs in a developing country: The case of Iran. International Journal of Technovation 28, 578–590 (2008)
10. Molla, A., Licker, P.: eCommerce adoption in developing countries: a model and instrument. International Journal of Information & Management 42, 877–899 (2005)
11. Kohli, A.K., Jaworski, B.J.: Market orientation: The construct research propositions, and managerial implications. International Journal of Marketing 54(2), 1–18 (1990)
12. Wu, F., Zsidisin, G.A., Ross, A.D.: Antecedents and Outcomes of E-procurement Adoption: An Integrative Model. IEEE Transactions on Engineering Management 54(3), 576–587 (2007)
13. Hui, L.Y.: An Empirical Investigation on the Determinants of E-procurement Adoption in Chinese Manufacturing Enterprises. In: International Conference on Management Science & Engineering (15th), Long Beach, USA, pp. 32–37 (2008)
14. Klein, K.J., Conn, A.B., Sorra, J.S.: Implementing Computerized Technology: An Organizational Analysis. International Journal of Applied Psychology 86(5), 811–824 (2001)
15. Iacovou, C., IBenbasat, I., Dexter, A.: Electronic data interchange and small organizations: adoption and impact of technology. MIS Quarterly 19(4), 465–485 (1995)
16. Lee, C.P., Lee, G.G., Lin, H.F.: The role of organizational capabilities in successful e-business implementation. International Journal of Business Process Management Journal 13(5), 677–693 (2007)
17. Rogers, E.M.: The Diffusion of Innovations. Free Press, New York (1995)
18. Tornatzky, L.G., Klein, K.J.: Innovation Characteristics and Innovation-Implementation: A Meta-Analysis of Findings. IEEE Transactions on Engineering Management 29(1), 28–45 (1982)
19. Tan, K.S., Eze, U.C.: An Empirical Study of Internet-Based ICT Adoption Among Malaysian SMEs. Communications of the IBIMA 1, 1–12 (2008)
20. Basu, A., Muylle, S.: Authentication in e-commerce. Communications of the ACM 46(12), 159–166 (2003)
21. Ahuja, M., Gupta, B., Raman, P.: An empirical investigation of online consumer purchasing behaviour. Communications of the ACM 46(12), 145–151 (2003)
22. Rahim, M.M.: Identifying Factors Affecting Acceptance of E-procurement Systems: An Initial Qualitative Study at an Australian City Council. Communications of the IBIMA 3, 7–17 (2008)
23. Kaliannan, M., Awang, H., Raman, M.: Government purchasing: A review of E-procurement system in Malaysia. The Journal of Knowledge Economy & Knowledge Management IV Spring (2009)
24. Chakravarthy, B.: A New Strategy Framework for Coping with Turbulence. In: Sloan Management Review, pp. 69–82 (1997)
25. Majumdar, S.K.: Sluggish giants, sticky cultures and dynamic capability transformation. Journal of Business Venturing 15(1), 59–78 (2000)
26. Pavlou, P.A.: IT-enabled dynamic capabilities in the new product development: Building a competitive advantage in the turbulent environments. Doctoral Dissertation, University of Southern California (2004)

# Author Index

Abran, Alain    42
Alves, Angela M.    145
Amengual, Esperança    64

Basri, Shuib Bin    240
Benthaus, Jens Peter    121
Bhandary, Smitha    186
Bharathi, V.    236
Bianco, Celestina    181

Casey, Valentine    73, 97, 208
Cawley, Oisín    84
Clarke, Paul    28
Coleman, Gerry    73

Demirors, Onur    108
Desharnais, Jean-Marc    42
Díaz, Rafael    169
Dominic, P.D.D.    240

Garzás, Javier    169

Halonen, Öjvind    52

Irrazabal, Emanuel    169

Johannessen, Per    52

Karasch, Timo    121
Keenan, Frank    217
Kelly, Sandra    217
Khan, Muhammad Jehangir    240

Laporte, Claude Y.    227
Lepmets, Marion    133

Mas, Antònia    64
McCaffery, Fergal    73, 97, 208, 217
McHugh, Martin    97
Mesquida, Antoni Lluís    64

Naseebullah,    240
Neumann, Robert    157

O'Connor, Rory V.    28, 227
O'Hara, Fran    194, 231
Örsmark, Ola    52

Pessoa, Marcelo    145
Plösch, Reinhold    221
Pomberger, Gustav    221

Quadri, Shah    186

Rajesh, G.    200
Ras, Eric    133
Rasking, Matthias    190
Renault, Alain    133
Richardson, Ita    84
Rout, Terry    1, 197

Salviano, Clênio F.    16, 145, 204
Schossleitner, Robert    157
Sechser, Bernhard    212
Shastry, Udaya    236
Sivakumar, M.S.    73
Sridhar, B.    200
Stallinger, Fritz    157, 221

Tarhan, Ayca    108
Tuffley, David    1

Vásquez, Felipe    169

Wang, Xiaofeng    84
Wen, Lian    1

Zarour, Mohammad    42
Zeilinger, Rene    157